Au

Questions and Answers for Electrician's Examinations

Audel™
Questions and Answers for Electrician's Examinations

All New Fourteenth Edition

Paul Rosenberg

WILEY

Wiley Publishing, Inc.

Vice President and Executive
 Publisher: Robert Ipsen
Vice President and Publisher:
 Joseph B. Wikert
Acquisitions Editor: Katie Feltman

Development Editor: Regina Brooks
Editorial Manager: Kathryn A. Malm
Production Editor: Vincent Kunkemueller
Text Design & Composition: TechBooks

Library of Congress Cataloging-in-Publication Data

Rosenberg, Paul.
  Audel questions and answers for electrician's examinations / Paul Rosenberg.— All new 14th ed.
    p. cm.
Rev. ed. of: Questions and answers for electricians examinations. 13th ed. 1999, revised by Paul Rosenberg.
ISBN: 0-7645-4201-X (PAPER/WEBSITE)
  1. Electric engineering—Examinations, questions, etc. 2. Electricians—Licenses—United States. 3. National Fire Protection Association. National Electrical Code (2002) I. Questions and answers for electricians examinations. II. Title.
  TK169.R67 2003
  621.319′24′076—dc22

                                        2003017921

Printed in the United States of America

10 9 8 7 6 5 4 3 2 1

Contents

Introduction

Tips on Taking Tests

It is the author's experience that, for most electricians, knowing how to take a test is almost as important as knowing the technical information, as far as obtaining a passing grade is concerned. A great number of electricians fear tests more than they fear 480 volts.

Really, there is no good reason why this should be so. After all, if hundreds of thousands of men and women can pass these tests, anyone interested who gives a real effort and pays particular attention to some basic rules can succeed. Some basic rules for taking tests are these:

1. Know the material being covered.
2. Know the format of the test.
3. Be physically and mentally prepared on the exam day.
4. RELAX!
5. Work the test the smartest way you can.

The first point—knowing the material being covered—is a mandatory prerequisite. Most test failures come from violating this rule. No, it isn't always easy to learn all the material on a test. It requires hours, sometimes many hours, of studying when you'd rather be doing other things. It means that you have to make your brain work harder than it wants to, going over the material again and again. Sorry, but unless you have an exceptional aptitude for learning, there are no shortcuts for hard, intense study. A good study guide (like this book) is about as much help as you can get.

The second rule for taking tests is that you need to *know the format of the test*. Some of the things you need to know are:

How many questions are on the test?

How many questions are open-book?

How many are closed-book?

Do all questions count for the same number of points?

Is there a penalty for wrong answers?

How much time is allowed for each section of the test?

Who wrote the test?

How will the test be graded?

By knowing the answers to these questions, you can plan your efforts intelligently. For example, if certain questions will count for more points than others, you should be ready to spend more time and effort on those questions. By knowing the time limits, you can calculate how much time you have for each question, etc. Get answers to all of these questions and consider all of these facts as you prepare for the exam.

Now, as for *being physically and mentally prepared*, I think most readers are familiar with the way athletes prepare for an event. They make sure they eat the right kinds of food so that they have enough energy. They get plenty of sleep, and they come to the event planning on winning. The same thing should be done in preparation for a test.

The most important factor is what was mentioned above—planning on winning. Psychologists have found that the results one achieves are directly related to what one expects to receive. If you believe that you will do well, you are quite likely to do well; if you believe that you will do poorly, you probably will. Remember, it does not matter what you wish for; what matters is what you actually expect to happen. I'll pass on to you one of my favorite quotes along these lines. It comes from Robert J. Ringer: "The results you produce in life are inversely proportional to the degree to which you are intimidated."

If you want to improve your confidence (expectations) in your test-taking abilities, picture yourself as having aced the test. Refuse to imagine yourself failing, and spend as much time studying as is necessary for you to believe in yourself.

On the day of a test, you want to walk in well rested (but not still groggy), having been well fed (but not full), and with a subdued confidence. Generally, heavy studying the night before the test is not a good idea. Do a light review and leisurely go over a difficult part of the information if you like, but the night before is not the time to get intense. You should have been intense two weeks ago. The night before the test is a time to eat well and to go to bed early. Try not to eat within two or three hours of the test, as it tends to bog you down. It has been said that mental efficiency is highest on an empty stomach.

Confidence is built on a good knowledge of the material to be covered and the ability to pass with style.

Upon entering the test location, *relaxing* is very important. If you choke up during the test, you are automatically taking five points off your score, possibly more. You should have the same attitude as the runner who shows up for a race he knows he will win. He is ready to run his fastest, but he is not nervous because he knows that his fastest is good enough.

Before taking the test, clear your mind, don't get involved with trivial conversations, and then, when it is time to answer the questions, dig into the test with your full strength.

During the test, first answer all of the easy questions; pass up the hard questions for now, and *do only the ones you know for sure.* Then, once you have answered these questions, don't go over them again; just move on to the next group of questions. Next, do the questions that will require some work, but don't do the most difficult questions; save them for last. It is silly to waste half of your time on one difficult question. Do the 47 easier questions, and then come back to the three especially difficult ones.

Work the test in the smartest possible way. Pay attention to time requirements, books allowed during open-book tests, etc. For your electrical exam, you should definitely put tabs on your Code book. Bring an electronic calculator with you and some scratch paper (as long as you are allowed to). Rather than buying a set of Code book tabs, I recommend that you do your own. Tab the index and the sections of the Code that you most commonly use. I generally put tabs on the following:

Tables 250.66 & 250.122 (Sizes of ground wires)

Table 310.16 (Wire ampacities)

Appendix C (Conduit fill tables)

Article 230 (Services)

Table 300.5 (Burial depths)

Table 370.6 (Number of wires in boxes)

Article 430 (Motors)

Article 450 (Transformers)

Article 490 (Over 600 volts)

Article 500 (Hazardous location wiring)

Article 700 (Emergency systems)

Remember
If hundreds of thousands of other people have passed these tests, you can too if you prepare.

Business Competency Examinations

In recent years, many municipalities have added business competency examinations to their standard Master Electrician examinations. In reality, they didn't have much choice. Since 1980, the number of licensed electrical contractors has skyrocketed, causing a great number of problems. Most of these problems were the result not of a lack of technical knowledge but of bad business practices. After some study, the various State Departments of Professional Regulation found out that while the newly licensed electrical contractors were proficient at trade skills they were woefully inadequate in business skills.

In an effort to ensure that newly licensed contractors are knowledgeable in business, new sections have been added to many competency examinations. Typically, 25 percent of a Master Electrician exam is dedicated to business skills and knowledge. The following are the topics usually covered:

1. Taxes
2. Unemployment and worker's compensation
3. OSHA and safety
4. Lien laws
5. Business skills

To help familiarize you with the various requirements and reference sources, each of these topics will be briefly discussed and then followed with questions and answers.

The *Taxes* section of such tests covers withholding of employee taxes. The information needed to answer these questions can be found in various IRS publications. (The easiest way to obtain these publications is to download them from the IRS's web site—http://www.irs.ustreas.gov/businesses/index.html.) Knowing the proper rules for withholding federal income tax, social security, federal unemployment, and state taxes is critical, not only for your test but also in order to operate a business. Let me state this clearly: The IRS is neither understanding nor compassionate, and it won't cut you even a little bit of slack for an ignorant infraction of its rules. The business of the IRS is to collect as much of your money as it is entitled to. Learn the rules for the test, and if and when you open a business, engage the services of a good accountant.

Unemployment compensation is paid directly to the state by the employer. It is not deducted from the employee's wages. Rates vary, and there are a number of requirements for anyone receiving this compensation. All of the required information can be found in a booklet called "Unemployment Compensation Handbook," which is available through various sources, including your public library.

Worker's compensation is handled on the state level, and the requirements vary from state to state; because of this, you will have to get local requirements from your state government. The people who administer your local test should be able to guide you to the right place.

OSHA (Occupational Safety and Health Administration) establishes rules to ensure that no employee is subjected to dangers to his or her safety or health. The OSHA regulations can be found in "OSHA Standard 2207, Part 1926." There are too many regulations to memorize, but one must be familiar enough with the book to be able to find the answer to any question easily.

Each state has its own *lien laws*. Copies of the regulations must be obtained through your own state government, although the testing agency administering your test can probably tell you exactly how to get them. Liens are very important in the construction business and have been developed primarily for the benefit of the contractor.

The *business skills* part of the test deals mostly with banks, financing, and basic management skills. As a reference source for the exam, "Tax Guide for Small Business" is recommended. This book, published by the IRS Division of the Treasury Department, is available from your local office of the Small Business Administration (SBA). There are many, many other business books available (and I would hope that anyone going into business would read several), but this handbook addresses the material in the test more directly.

You should remember, however, that the business skills covered by these tests are not enough to ensure success in business. In addition to these skills, you will need skill in dealing with people, the ability to analyze a market, and the ability to make and follow through on decisions. This test covers only academic business skills; to actually make money, you will need other skills also.

I-1   If a certain employee spends less than half of his time during a pay period performing services that are subject to taxation, how much of his or her pay is taxable?
All employees are taxable.

I-2   If an employer fails to make federal income tax deposits when they are due, how large a penalty will they be assessed?
10 percent.

I-3 A self-employed person is considered an employee. True or false?
   False.

I-4 What form must be used to correct errors in withholding taxes?
   941C.

I-5 What would be the take-home pay of a worker who claims one deduction, is married, and who earns $500.00 per week (without state or local taxes)?

500.00	
(56.00)	Federal Income Tax Withholding
(31.00)	Social Security
(7.25)	Medicare
$405.75	

I-6 For unemployment taxation, the term "employer" includes any person or organization that paid _____ or more in wages in any quarter or had employees at any time in 20 weeks of the year.
   $1,500.00.

I-7 If an employee is paid $325.00 per week, how much Social Security and Medicare tax should be deducted from his or her wages?
   $325.00 \times 7.65\% = \$24.86$.

I-8 What form is used to get an Employer Identification Number?
   SS-4.

I-9 What form must a new employee sign before beginning work?
   W-4.

I-10 If an employer has _____ or more employees in 20 or more weeks, the employer must file a Form 940 Federal Unemployment Act.
   1.

I-11 On what portion of his or her wages must an employee contribute for state unemployment compensation?
   The first $9,000.00 of wages.

I-12 Casual labor is labor that is occasional, incidental, and not exceeding ___ working days in duration.
  10.

I-13 Does a temporary light fixture with a reflector that deeply recesses its bulb require a guard?
  No.

I-14 How long may double cleat ladders be?
  15 feet.

I-15 What is the angle of repose for average soil?
  45 degrees.

I-16 How many gallons of flammable liquid can be stored in a room outside of an approved storage cabinet?
  25.

I-17 When an interior-hung scaffold is suspended from the beams of a ceiling, what percentage of the rated load must the suspending wire be capable of supporting?
  600 percent (six times the rated load).

I-18 What is the proper maintenance procedure for an "ABC" dry-chemical stored-pressure fire extinguisher?
  Check the pressure gauge and the condition of the chemical annually.

I-19 Workers should not be exposed to impulsive or impact noises louder than _____ decibels.
  140.

I-20 What is the standard height for a guardrail?
  42 inches.

I-21 What is the minimum size (OSHA requirement) of a conductor to a ground rod?
  #2 AWG copper.

I-22 For a scaffold with a working load of 75 pounds per square feet, what is the maximum span for a 2″ × 12″ plank?
  7 feet.

I 23 Loaded powder activated tools may not be left _____
  Unattended.

I-24 What should be the predominant color of caution signs?
Yellow.

I-25 For 225 employees on a construction site, how many toilets must be provided?
Five toilet seats and five urinals.

I-26 When safety belts are used, the maximum distance of fall must be _____.
6 feet.

I-27 What is the minimum lighting level in a field construction office?
30 foot-candles.

I-28 Manually handled lumber cannot be stacked higher than _____.
16 feet.

I-29 What is the term for the claims of a creditor against the assets of a business?
Liabilities.

I-30 Small Business Administration loans can be guaranteed up to _____.
90 percent.

I-31 Would taxes be considered a liability?
Yes.

I-32 The assets of a business, minus its liabilities, are called its _____.
Equity.

I-33 Accounts receivable financing is normally based on receivables that are how old?
70 to 90 days.

I-34 If your company has gross sales of $210,000.00 and expenses of $198,500.00, what percentage of profit did it make?
5.5 percent.

I-35 An agreement by which you get exclusive use of a certain item for a stated period of time is called a _____.
Lease.

I-36 What is the minimum rate of return sought by most venture capitalists?

It is currently around 10 percent (or higher), but this figure can be modified by changes in interest rates, inflation, etc.

I-37 A one-year line of credit refers to a note that is renewable for one year at ___-day intervals.

90.

I-38 What is the term for a plan of cash receipts and expenditures for a certain period of time?

A cash budget.

I-39 What is the term for the money required to carry accounts receivable, cover payrolls, and buy products?

Working capital.

I-40 Is an 18-year-old boy, employed by his parents, exempt from Social Security tax?

Yes.

I-41 Is a wife, employed by her husband, subject to Social Security tax?

No.

I-42 What are the two primary account methods?

Cash accounting and accrual.

I-43 Are all types of business activities voluntary?

No, the payment of taxes is enforced. Almost every other type of business activity is voluntary between the parties involved.

I-44 What type of law covers the awarding of damages for accidental injuries and the like?

Torts.

I-45 Define "overhead."

"Overhead" is the money necessary to keep a company operating, even if there is no one working in the field. It includes everything except material, labor, and job expenses. Office expenses, office salaries, sales expenses, office equipment, vehicles, and similar expenses are considered to be overhead.

I-46 What does the term "Net 30" indicate?

"Net 30" indicates a payment procedure. In general, it means that if one party to a transaction presents a valid invoice, it will be paid by the other party within 30 days.

I-47 What is "cash flow," and why is it important?

"Cash flow" is a general term describing the flow of cash within a company. It is important because electrical construction work is almost always done on credit, sometimes leaving the contractor in a situation in which he or she is making a lot of money but hasn't collected it yet and therefore has no cash with which to pay bills. Insufficient cash has been the ruin of many construction firms.

I-48 What are the functions of profit in a company?

There are two. The first is to offset risks. Without some extra money in a contract, even a small difficulty would cause the project to go over budget. The second is to give the owners of the company a return on their investment. If the owners did not get a return on their money, they would have no reason to put it to use in the company.

I-49 What are job expenses?

Expenses—such as storage trailer rental, tool rentals, and job telephones—that are caused by the project and not by continuing operations.

I-50 How does OSHA make sure that their rules are followed?

By imposing fines on companies that are judged to be in violation.

Electrical Symbols

To avoid confusion, ASA policy requires that the same symbol not be included in more than one Standard. If the same symbol were used in two or more Standards and one of these Standards were revised, changing the meaning of the symbol, considerable confusion could arise over which symbol was correct, the revised or unrevised.

The symbols in this category include, but are not limited to, those listed below. The reference numbers are the American Standard Y32.2 item numbers.

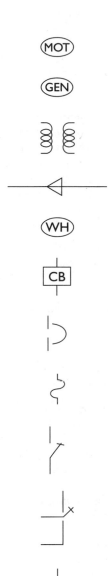

46.3   *Electric motor*

46.2   *Electric generator*

86.1   *Power transformer*

82.1   *Pothead (cable termination)*

48   *Electric watthour meter*

12.2   *Circuit element, e.g.,
          circuit breaker*

11.1   *Circuit breaker*

36   *Fusible element*

76.3   *Single-throw knife switch*

76.2   *Double-throw knife switch*

13.1   *Ground*

7   *Battery*

Electrical Symbols.

List of Symbols
1.0 Lighting Outlets

Ceiling   Wall

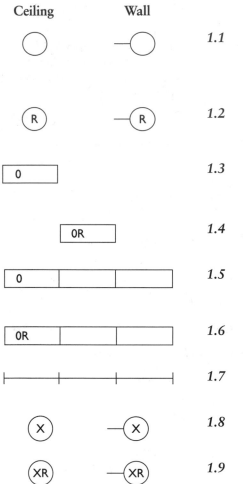

1.1   *Surface or pendant incandescent, mercury vapor, or similar lamp fixture*

1.2   *Recessed incandescent, mercury vapor, or similar lamp fixture*

1.3   *Surface or pendant individual fluorescent fixture*

1.4   *Recessed individual fluorescent fixture*

1.5   *Surface or pendant continuous-row fluorescent fixture*

1.6   *Recessed continuous-row fluorescent fixture**

1.7   *Bare-lamp fluorescent strip***

1.8   *Surface or pendant exit light*

1.9   *Recessed exit light*

(continued)

---

*In the case of combination continuous-row fluorescent and incandescent spotlights, use combinations of the above Standard symbols.

**In the case of a continuous-row bare-lamp fluorescent strip above an area-wide diffusion means, show each fixture run, using the Standard symbol; indicate area of diffusing means and type of light shading and/or drawing notation.

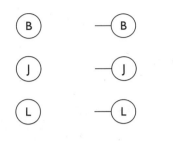

1.10   *Blanked outlet*

1.11   *Junction box*

1.12   *Outlet controlled by low-voltage switching when relay is installed in outlet box*

Lighting Outlets.

2.0 Receptacle Outlets

Unless noted to the contrary, it should be assumed that every receptacle will be grounded and will have a separate grounding contact.

Use the uppercase subscript letters described under Section 2 item a-2 of this Standard when weatherproof, explosion-proof, or some other specific type of device will be required.

2.1   *Single receptacle outlet*

2.2   *Duplex receptacle outlet*

2.3   *Triplex receptacle outlet*

2.4   *Quadruplex receptacle outlet*

2.5   *Duplex receptacle outlet—split wired*

2.6   *Triplex receptacle outlet—split wired*

2.7   *Single special-purpose receptacle outlet**

(continued)

---

*Use numeral or letter, either within the symbol or as a subscript alongside the symbol keyed to explanation in the drawing list of symbols, to indicate type of receptacle or usage.

2.8 *Duplex special-purpose receptacle outlet**

2.9 *Range outlet*

2.10 *Special-purpose connection or provision for connection. Use subscript letters to indicate function (DW— dishwasher; CD—clothes dryer, etc.)*

2.11 *Multioutlet assembly. Extend arrows to limit of installation. Use appropriate symbol to indicate type of outlet. Also indicate spacing of outlets as x inches.*

2.12 *Clock Hanger Receptacle*

2.13 *Fan Hanger Receptacle*

2.14 *Floor Single Receptacle Outlet*

2.15 *Floor Duplex Receptacle Outlet*

2.16 *Floor Special-Purpose Outlet**

(continued)

---

*Use numeral or letter, either within the symbol or as a subscript alongside the symbol keyed to explanation in the drawing list of symbols, to indicate type of receptacle or usage.

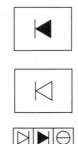

2.17   Floor Telephone
       Outlet—Public

2.18   Floor Telephone
       Outlet—Private

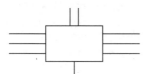

*Not a part of the Standard:
example of the use of several
floor outlet symbols to identify
a 2-, 3-, or more-gang floor
outlet*

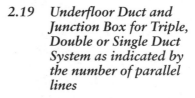

2.19   Underfloor Duct and
       Junction Box for Triple,
       Double or Single Duct
       System as indicated by
       the number of parallel
       lines

*Not a part of the Standard:
example of use of various
symbols to identify location of
different types of outlets or
connections for underfloor duct
or cellular floor systems*

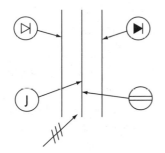

2.20   Cellular Floor Header
       Duct

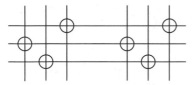

Receptacle Outlets.

3.0 Switch Outlets

S	*3.1*	*Single-pole switch*	
S_2	*3.2*	*Double-pole switch*	
S_3	*3.3*	*Three-way switch*	
S_4	*3.4*	*Four-way switch*	
S_K	*3.5*	*Key-operated switch*	
S_P	*3.6*	*Switch and pilot lamp*	
S_L	*3.7*	*Switch for low-voltage switching system*	
S_{LM}	*3.8*	*Master switch for low-voltage switching system*	

$-\!\!\ominus\, S$　　*3.9*　　*Switch and single receptacle*

$=\!\!\ominus\, S$　　*3.10*　　*Switch and double receptacle*

S_D	*3.11*	*Door switch*	
S_T	*3.12*	*Time switch*	
S_{CB}	*3.13*	*Circuit-breaker switch*	
S_{MC}	*3.14*	*Momentary contact switch or pushbutton for other than signaling system*	

Switch Outlets.

Signaling System Outlets

4.0 Institutional, Commercial, and Industrial Occupancies

These symbols are recommended by the American Standards Association but are not used universally. The reader should remember not to assume that these symbols will be used on any certain plan and should always check the symbol list on the plans to verify whether these symbols are actually used.

Basic
Symbol

Examples of
Individual Item
Identification
(Not a part of
the Standard)

4.1 I.   Nurse Call System Devices
          (and type)

*Nurses' Annunciator (can add a
number after it as* ⊕ *24 to
indicate number of lamps)*

*Call station, single cord, pilot
light*

*Call station, double cord,
microphone speaker*

Corridor dome light, 1 lamp

Transformer

*Any other item on same system—
use numbers as required.*

4.2 II. Paging System Devices
          (any type)

Keyboard

Flush annunciator

2-face annunciator

*Any other item on same system—
use numbers as required.*

(continued)

Basic
Symbol

Examples of
Individual Item
Identification
(Not a part of
the Standard)

4.3 III. Fire Alarm System Devices
(any type) including Smoke
and Sprinkler Alarm Devices

 Control panel

Station

 10" Gong

 Pre-signal chime

 *Any other item on same system—
use numbers as required.*

4.4 IV. Staff Register System Devices
(any type)

Phone operators' register

Entrance register—flush

Staff room register

Transformer

 *Any other item on same system—
use numbers as required.*

(continued)

Basic Symbol	Examples of Individual Item Identification (Not a part of the Standard)

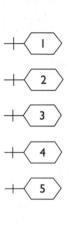

4.5 V. Electric Clock System Devices (any type)

Master clock

12" Secondary—flush

12" Double dial—wall-mounted

18" Skeleton dial

Any other item on same system—use numbers as required.

4.6 VI. Public Telephone System Devices

Switchboard

Desk phone

Any other item on same system—use numbers as required.

4.7 VII. Private Telephone System Devices (any type)

Switchboard

Wall phone

Any other item on same system—use numbers as required.

(continued)

Basic Symbol	Examples of Individual Item Identification (Not a part of the Standard)		

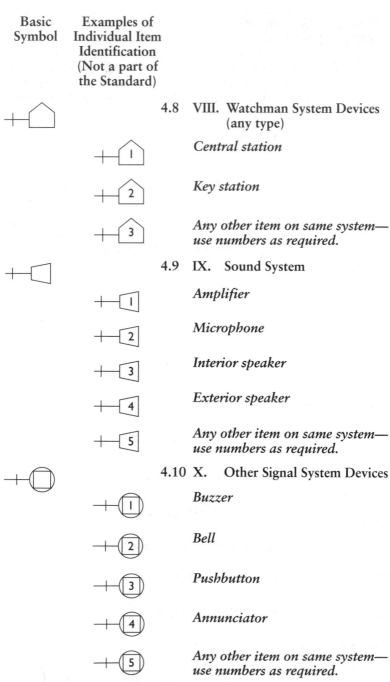

4.8 VIII. Watchman System Devices (any type)

Central station

Key station

Any other item on same system— use numbers as required.

4.9 IX. Sound System

Amplifier

Microphone

Interior speaker

Exterior speaker

Any other item on same system— use numbers as required.

4.10 X. Other Signal System Devices

Buzzer

Bell

Pushbutton

Annunciator

Any other item on same system— use numbers as required.

Institutional, Commercial, and Industrial Occupancies.

Signaling System Outlets

5.0 Residential Occupancies

When a descriptive symbol list is not employed, use the following signaling system symbols to identify standardized, residential-type, signal-system items on residential drawings. Use the basic symbols with a descriptive symbol list when other signal-system items are to be identified.

Symbol		
	5.1	*Pushbutton*
	5.2	*Buzzer*
	5.3	*Bell*
	5.4	*Combination bell-buzzer*
CH	5.5	*Chime*
	5.6	*Annunciator*
D	5.7	*Electric door opener*
M	5.8	*Maid's signal plug*
	5.9	*Interconnection box*
BT	5.10	*Bell-ringing transformer*
	5.11	*Outside telephone*
	5.12	*Interconnecting telephone*
R	5.13	*Radio outlet*
TV	5.14	*Television outlet*

Residential Occupancies.

6.0 Panelboards, Switchboards, and Related Equipment

6.1 *Flush-mounted panelboard and cabinet**

6.2 *Surface-mounted panelboard and cabinet**

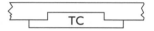

6.3 *Switchboard, power control center, unit substations*— should be drawn to scale*

6.4 *Flush-mounted terminal cabinet.* In small-scale drawings the TC may be indicated alongside the symbol.*

6.5 *Surface-mounted terminal cabinet.* In small-scale drawings the TC may be indicated alongside the symbol.*

6.6 *Pull box (identify in relation to wiring section and sizes)*

6.7 *Motor or other power controller**

6.8 *Externally-operated disconnection switch**

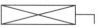

6.9 *Combination controller and disconnection means**

Panelboards, Switchboards, and Related Equipment.

---

*Identify by notation or schedule.

7.0 Bus Ducts and Wireways

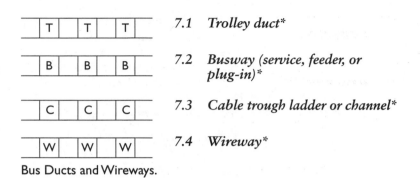

7.1 *Trolley duct**

7.2 *Busway (service, feeder, or plug-in)**

7.3 *Cable trough ladder or channel**

7.4 *Wireway**

Bus Ducts and Wireways.

8.0 Remote Control Stations for Motors or Other Equipment*

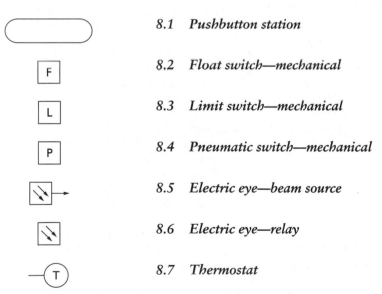

8.1 *Pushbutton station*

8.2 *Float switch—mechanical*

8.3 *Limit switch—mechanical*

8.4 *Pneumatic switch—mechanical*

8.5 *Electric eye—beam source*

8.6 *Electric eye—relay*

8.7 *Thermostat*

Remote Control Stations for Motor or Other Equipment.

---

*Identify by notation or schedule.

9.0 Circuiting
Wiring method identification by notation on drawing or in specification.

9.1 *Wiring concealed in ceiling or wall*

9.2 *Wiring concealed in floor*

9.3 *Wiring exposed*

Note: *Use heavyweight line to identify service and feeders. Indicate empty conduit by notation CO (conduit only).*

2   1

3 wires

9.4 *Branch-circuit home run to panel-board. Number of arrows indicates number of circuits. (A numeral at each arrow may be used to identify circuit number.) Note: Any circuit without further identification indicates two-wire circuit. For a greater number of wires, indicate with cross lines, e.g.:*

4 wires, etc.

Unless indicated otherwise, the wire size of the circuit is the minimum size required by the specification.

Identify different functions of wiring system, e.g., signaling system by notation or other means.

9.5 *Wiring turned up*

9.6 *Wiring turned down*

Circuiting.

10.0 Electric Distribution or Lighting System, Underground

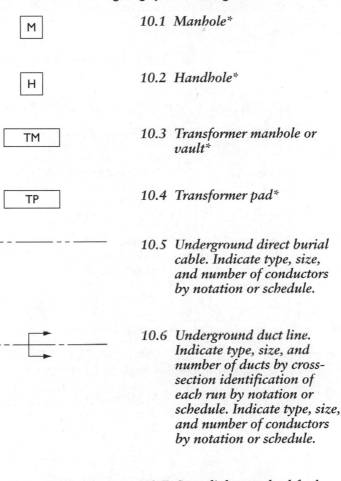

10.1 *Manhole**

10.2 *Handhole**

10.3 *Transformer manhole or vault**

10.4 *Transformer pad**

10.5 *Underground direct burial cable. Indicate type, size, and number of conductors by notation or schedule.*

10.6 *Underground duct line. Indicate type, size, and number of ducts by cross-section identification of each run by notation or schedule. Indicate type, size, and number of conductors by notation or schedule.*

10.7 *Streetlight standard feed from underground circuit**

Electric Distribution or Lighting System, Underground.

---

*Identify by notation or schedule.

11.0 Electric Distribution or Lighting System, Aerial

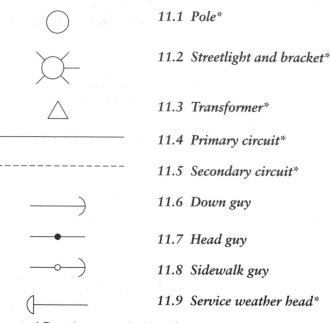

*11.1 Pole**

*11.2 Streetlight and bracket**

*11.3 Transformer**

*11.4 Primary circuit**

*11.5 Secondary circuit**

11.6 Down guy

11.7 Head guy

11.8 Sidewalk guy

*11.9 Service weather head**

Electrical Distribution or Lighting System Aerial.

4 Arrester, Lighting Arrester (Electric surge, etc.) Gap

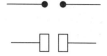

4.1   General

4.2   Carbon block

Block, telephone protector

*The sides of the rectangle are
to be approximately in the
ratio of 1 to 2, and the space
between rectangles shall be
approximately equal to the
width of a rectangle.*

(continued)

---

*Identify by notation or schedule.

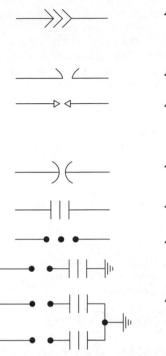

4.3 *Electrolytic or aluminum cell*
     This symbol is not
     composed of arrowheads.

4.4 *Horn gap*

4.5 *Protective gap*
     These triangles shall not be
     filled.

4.6 *Sphere gap*

4.7 *Valve or film element*

4.8 *Multigap, general*

4.9 *Application: gap plus valve*
     plus ground, 2-pole4.9
     Application: gap plus valve
     plus ground, 2-pole

Arrester, Lighting Arrester (Electric surge etc.) Gap.

7 Battery

The long line is always positive,
but polarity may be indicated in
addition. Example:

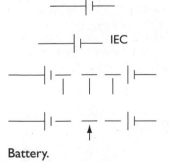

7.1 *Generalized direct-current*
     source

7.2 *One cell*

7.3 *Multicell*

7.3.1 *Multicell battery with 3 taps*

7.3.2 *Multicell battery with*
       adjustable tap

Battery.

11 Circuit Breakers

If it is desired to show the condition causing the breaker to trip, the relay-protective-function symbols in item 66.6 may be used alongside the break symbol.

IEC

11.1 *General*

IEC

11.2 *Air circuit breaker, if distinction is needed; for alternating-current breakers rated at 1,500 volts or less and for all direct-current circuit breakers.*

11.2.1 *Network protector*

IEC   IEC

See Note 11.3A

11.3 *Circuit breaker, other than covered by item 11.2. The symbol in the right column is for a 3-pole breaker.*

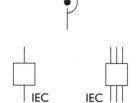

See Note 11.3A

11.3.1 *On a connection or wiring diagram, a 3-pole single-throw circuit breaker (with terminals shown) may be drawn as shown.*

(continued)

Note 11.3A—On a power diagram, the symbol may be used without other identification. On a composite drawing where confusion with the general circuit element symbol (item 12) may result, add the identifying letters CB inside or adjacent to the square.

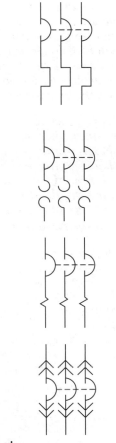

11.4 Applications

11.4.1 3-pole circuit breaker with thermal overload device in all 3 poles

11.4.2 3-pole circuit breaker with magnetic overload device in all 3 poles

11.4.3 3-pole circuit breaker, drawout type

Circuit Breakers.

13 Circuit Return

IEC

13.1 Ground

(A) A direct conducting connection to the earth or body of water that is a part thereof

(continued)

(B) *A conducting connection to a structure that serves a function similar to that of an earth ground (that is, a structure such as a frame of an air, space, or land vehicle that is not conductively connected to earth)*

IEC

13.2 *Chassis or frame connection*
A conducting connection to a chassis or frame of a unit. The chassis or frame may be at a substantial potential with respect to the earth or structure in which this chassis or frame is mounted.

13.3 *Common connections*
Conducting connections made to one another. All like-designated points are connected. *The asterisk is not a part of the symbol. Identifying valves, letters, numbers, or marks shall replace the asterisk.

*

⧖*

Circuit Return.

15 Coil, Magnetic Blowout*

Coil, Magnetic Blowout.

---

*The broken line (— - —) indicates where line connection to a symbol is made and is not a part of the symbol.

23 Contact, Electrical

For buildups or forms using electrical contacts, see applications under CONNECTOR (item 19), RELAY (item 66), and SWITCH (item 76). See DRAFTING PRACTICES (item 0.4.6).

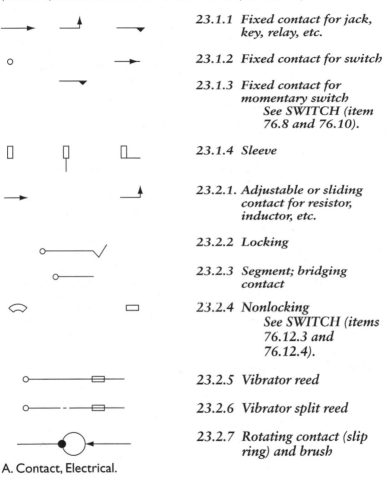

23.1.1 *Fixed contact for jack, key, relay, etc.*

23.1.2 *Fixed contact for switch*

23.1.3 *Fixed contact for momentary switch*
See SWITCH (item 76.8 and 76.10).

23.1.4 *Sleeve*

23.2.1. *Adjustable or sliding contact for resistor, inductor, etc.*

23.2.2 *Locking*

23.2.3 *Segment; bridging contact*

23.2.4 *Nonlocking*
See SWITCH (items 76.12.3 and 76.12.4).

23.2.5 *Vibrator reed*

23.2.6 *Vibrator split reed*

23.2.7 *Rotating contact (slip ring) and brush*

A. Contact, Electrical.

It is standard procedure to show a contact by a symbol that indicates the circuit condition produced when the actuating device is in the nonoperated, or deenergized, position. It may be necessary to add a clarifying note explaining the proper point at which the contact functions — the point where the actuating device (mechanical, electrical, etc.) opens or closes due to changes in pressure, level, flow, voltage, current, etc. When it is necessary to show contacts in the operated, or energized, condition—and where confusion would otherwise result—a clarifying note shall be added to the drawing. Contacts for circuit breakers, auxiliary switches, etc., may be designated as shown below:

(a) Closed when device is in energized or operated position.

(b) Closed when device is in deenergized or nonoperated position.

(aa) Closed when operating mechanism of main device is in energized or operated position.

(bb) Closed when operating mechanism of main device is in deenergized or nonoperated position.

[See American Standard C37.2-1962 for details.]

In the parallel-line contact, symbols showing the length of the parallel lines shall be approximately 1¼ times the width of the gap (except for item 23.6).

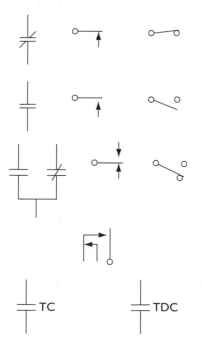

23.3.1 Closed contact (break)
See also SWITCHING
FUNCTION (item 77).

23.3.2 Open contact (make)
See also SWITCHING
FUNCTION (item 77).

23.3.3 Transfer
See also SWITCHING
FUNCTION (item 77).

23.3.4 Make-before-break

23.4   Application: open contact
with time closing (TC) or
time delay closing (TDC)
feature

(continued)

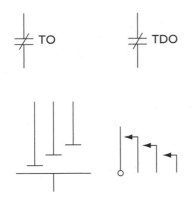

23.5   Application: closed contact with time opening (TO) or time delay opening (TDO) feature

23.6   Time sequential closing

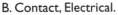

B. Contact, Electrical.

24 Contactor
See also RELAY (item 66).

Contactor symbols are derived from fundamental contact, coil, and mechanical connection symbols and should be employed to show contactors on complete diagrams. A complete diagram of the actual contactor device is constructed by combining the abovementioned fundamental symbols for mechanical connections, control circuits, etc.

Mechanical interlocking should be indicated by notes.

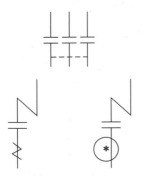

24.1   Manually operated 3-pole contactor

24.2   Electrically operated 1-pole contact or with series blowout coil See Note 24.2A.

(continued)

Note 24.2A   The asterisk is not a part of the symbol. Always replace the asterisk by a device designation.

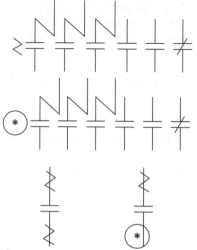

24.3 *Electrically operated 3-pole contactor with series blowout coils; 2 open and 1 closed auxiliary contacts (shown smaller than the main contacts)*

24.4 *Electrically operated 1-pole contactor with shunt blowout coil*

Contactor.

46 Machine, Rotating

46.1 *Basic*

46.2 *Generator (general)*

46.3 *Motor (general)*

46.4 *Motor, multispeed*

USE BASIC MOTOR SYMBOL AND
NOTE SPEEDS

46.5 *Rotating armature with commutator and brushes**

(continued)

---

*The broken line (–––) indicates where line connection to a symbol is made and is not a part of the symbol.

46.6 *Field, generator or motor*
Either symbol of item 42.1
may be used in the following
items.

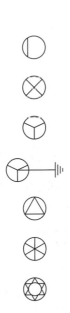

46.6.1 *Compensating or commutating*

46.6.2 *Series*

46.6.3 *Shunt, or separately excited*

46.6.4 *Magnet, permanent*
See item 47.

46.7 *Winding symbols*
Motor and generator
winding symbols may be
shown in the basic circle
using the following
representation.

46.7.1 *1-phase*

46.7.2 *2-phase*

46.7.3 *3-phase wye (ungrounded)*

46.7.4 *3-phase wye (grounded)*

46.7.5 *3-phase delta*

46.7.6 *6-phase diametrical*

46.7.7 *6-phase double-delta*

46.8 *Direct-current machines;*
applications

(continued)

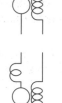

46.8.1 *Separately excited direct-current generator or motor**

46.8.2 *Separately excited direct-current generator or motor; with commutating or compensating field winding or both**

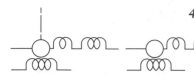

46.8.3 *Compositely excited direct-current generator or motor; with commutating or compensating field winding or both**

46.8.4 *Direct-current series motor or 2-wire generator**

46.8.5 *Direct-current series motor or 2-wire generator; with commutating or compensating field winding or both**

46.8.6 *Direct-current shunt motor or 2-wire generator**

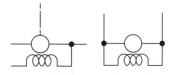

46.8.7 *Direct-current shunt motor or 2-wire generator; with commutating or compensating field winding or both**

(continued)

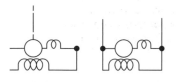

*The broken line (- — -) indicates where line connection to a symbol is made and is not a part of the symbol.

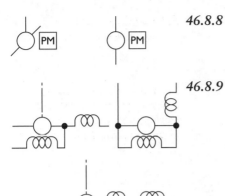

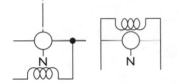

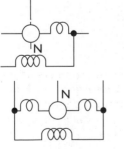

46.8.8 *Direct-current, permanent-magnet field generator or motor**

46.8.9 *Direct-current, compound motor or 2-wire generator or stabilized shunt motor**

46.8.10 *Direct-current compound motor or 2-wire generator or stabilized shunt motor; with commutating or compensating field winding or both**

46.8.11 *Direct-current, 3-wire shunt generator**

46.8.12 *Direct-current, 3-wire shunt generator; with commutating or compensating field winding or both**

(continued)

"The broken line (- -) indicates where line connection to a symbol is made and is not a part of the symbol.

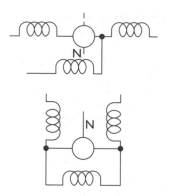

46.8.13 *Direct-current, 3-wire compound generator**

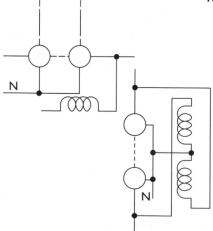

46.8.14 *Direct-current, 3-wire compound generator; with commutating or compensating field winding or both**

46.8.15 *Direct-current balancer, shunt wound**

(continued)

*The broken line (- — -) indicates where line connection to a symbol is made and is not a part of the symbol.

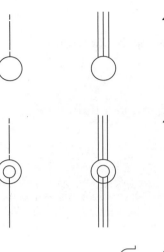

46.9 Alternating-current machines; application

46.9.1 Squirrel-cage induction motor or generator, split-phase induction motor or generator, rotary phase converter or repulsion motor*.

46.9.2 Wound-rotor induction motor, synchronous induction motor, induction generator, or induction frequency converter*

46.9.3 Alternating-current series motor*

Machine, Rotating.

48 Meter Instrument

As indicated in Note 48A, the asterisk is not part of the symbol and should always be replaced with one of the letter combinations listed below, according to the meter's function. This is not necessary if some other identification is provided in the circle and described in the diagram.

A   Ammeter

AH   Ampere-hour

CMA   Contact-making (or breaking) ammeter

(continued)

---

*The broken line (▪ ━ ▪) indicates where line connection to a symbol is made and is not a part of the symbol.

CMC	*Contact-making (or breaking) clock*
CMV	*Contact-making (or breaking) voltmeter*
CRO	*Oscilloscope or cathode-ray oscillograph*
DB	*DB (decibel) meter*
DBM	*DBM (decibels referred to 1 milliwatt) meter*
DM	*Demand meter*
DTR	*Demand-totalizing relay*
F	*Frequency meter*
G	*Galvanometer*
GD	*Ground detector*
I	*Indicating*
INT	*Integrating*
μA or UA	*Microammeter*
MA	*Milliammeter*
NM	*Noise meter*
OHM	*Ohmmeter*
OP	*Oil pressure*
OSCG	*Oscillograph string*

(continued)

PH	*Phasemeter*
PI	*Position indicator*
PF	*Power factor*
RD	*Recording demand meter*
REC	*Recording*
RF	*Reaction factor*
SY	*Synchroscope*
TLM	*Telemeter*
T	*Temperature meter*
THC	*Thermal converter*
TT	*Total time*
V	*Voltmeter*
VA	*Volt-ammeter*
VAR	*Varmeter*
VARH	*Varhour meter*
VI	*Volume indicator; meter, audio level*
VU	*Standard volume indicator; meter, audio level*
W	*Wattmeter*
WH	*Watthour meter*

Meter Instrument.

58 Path, Transmission, Conductor, Cable, Wiring

————————————

58.1   *Guided path, general*
The entire group of
conductors, or the
transmission path
required to guide the
power or symbol, is
shown by a single line.
In coaxial and
waveguide work, the
recognition symbol is
employed at the
beginning and end of
each type of transmission
path as well as at
intermediate points to
clarify a potentially
confusing diagram. For
waveguide work, the
mode may be indicated   ·
as well.

————————————

58.2   *Conductive path or*
       conductor; wire

58.2.1 *Two conductors or*
       conductive paths of wires

58.2.2 *Three conductors or*
       conductive paths of wires

58.2.3 *"n" conductors or*
       conductive paths of wires

58.5   *Crossing of paths or*
       conductors not
       connected
          The crossing is not
          necessarily at a 90-
          degree angle.

(continued)

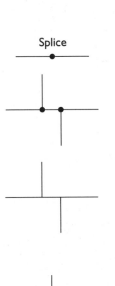

58.6   *Junction of paths or conductors*

58.6.1   *Junction of paths or conductors*

58.6.1.1 *Application: junction of paths, conductor, or cable. If desired, indicate path type or size.*

58.6.1.2 *Application: splice (if desired) of same size cables. Junction of conductors of same size or different size cables. If desired, indicate sizes of conductors.*

58.6.2   *Junction of connected paths, conductors, or wires*

OR ONLY IF REQUIRED BY SPACE LIMITATION

Path, Transmission, Conductor, Cable, Wiring.

63 Polarity Symbol

+   63.1   *Positive*

–   63.2   *Negative*

Polarity Symbol.

76 Switch

See also FUSE (item 36); CONTACT, ELECTRIC (item 23); and DRAFTING PRACTICES (items 0.4.6 and 0.4.7).

Switch symbols may be constructed using the fundamental symbols for mechanical connections, contacts, etc.

In standard procedure, a switch is represented in the nonoperating, or deenergized, position. In the case of switches that have two or more positions in which no operating force is applied and for those switches (air-pressure, liquid-level, rate-of-flow, etc.) that may be actuated by a mechanical force, the point at which the switch functions should be described in a clarifying note.

In cases where the basic switch symbols (items 76.1–76.4) are used in a diagram in the closed position, the terminals must be included for clarity.

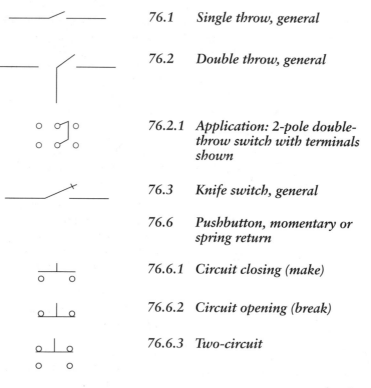

76.1   *Single throw, general*

76.2   *Double throw, general*

76.2.1   *Application: 2-pole double-throw switch with terminals shown*

76.3   *Knife switch, general*

76.6   *Pushbutton, momentary or spring return*

76.6.1   *Circuit closing (make)*

76.6.2   *Circuit opening (break)*

76.6.3   *Two-circuit*

(continued)

76.7   *Pushbutton, maintained or not*
        spring return

76.7.1 *Two-circuit*

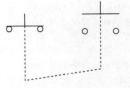

Switch.

86 Transformer

86.1   *General*

*Either winding symbol may be used.
In the following items, the left symbol
is used. Additional windings may be
shown or indicated by a note. For
power transformers use polarity
markingH$_1$, X$_1$, etc., from American
Standard C6.1-1956.*

*For polarity markings on current and potential
transformers, see items 86.16.1 and 86.17.1*

*In coaxial and waveguide circuits , this symbol will
represent a taper or step transformer without mode
change*

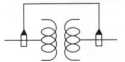

86.1.1 *Application: transformer
        with direct-current
        connections and mode
        suppression between two
        rectangular waveguides*

86.2   *If it is desired especially to
        distinguish a magnetic-core
        transformer*

86.2.1 *Application: shielded
        transformer with magnetic
        core shown*

(continued)

86.2.2 *Application: transformer with magnetic core shown and with a shield between windings. The shield is shown connected to the frame.*

86.6 *With taps, 1-phase*

86.7 *Autotransformer, 1-phase*

86.7.1 *Adjustable*

86.13 *1-phase, 2-winding transformer*

86.13.1 *3-phase bank of 1-phase, 2-winding transformer*

See American Standard C6.1-1965 for interconnections for complete symbol.

86.14 *Polyphase transformer*

(continued)

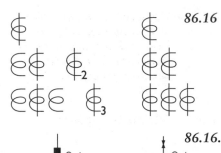

86.16   Current transformer(s)

86.16.1 Current transformer with polarity marking. *Instantaneous direction of current into one polarity mark corresponds to current out of the other polarity mark.*

Symbol used shall not conflict with item 77 when used on same drawing.

86.16.2 Bushing-type current transformer*

86.17   Potential transformer(s)

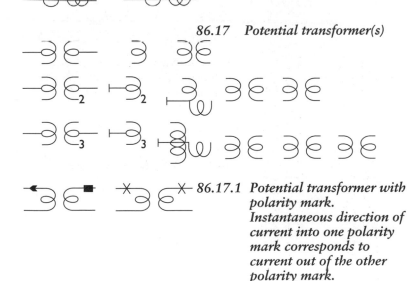

86.17.1 Potential transformer with polarity mark. *Instantaneous direction of current into one polarity mark corresponds to current out of the other polarity mark.*

(continued)

---

*The broken line (- — -) indicates where line connection to a symbol is made and is not a part of the symbol.

Symbol used shall not conflict with item 77 when used on same drawing.

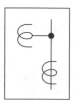

86.18   *Outdoor metering device*

86.19   *Transformer winding connection symbols*

For use adjacent to the symbols for the transformer windings.

86.19.1   *2-phase 3-wire, grounded*

86.19.1.1   *2-phase 3-wire, grounded*

86.19.2   *2-phase 4-wire*

86.19.2.1   *2-phase 5-wire, grounded*

86.19.3   *3-phase 3-wire, delta or mesh*

86.19.3.1   *3-phase 3-wire, delta, grounded*

86.19.4   *3-phase 4-wire, delta, ungrounded*

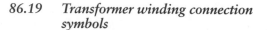

(continued)

86.19.4.1 *3-phase 4-wire, delta, grounded*

86.19.5 *3-phase, open-delta*

86.19.5.1 *3-phase, open-delta, grounded at common point*

86.19.5.2 *3-phase, open-delta, grounded at middle point of one transformer*

86.19.6 *3-phase, broken-delta*

86.19.7 *3-phase, wye or star, ungrounded*

86.19.7.1 *3-phase, wye, grounded neutral*

The direction of the stroke representing the neutral can be arbitrarily chosen.

86.19.8 *3-phase 4-wire, ungrounded*

Transformer.

Chapter 1

Review Definitions

Definitions are covered in Article 100 of the NEC (National Electrical Code). The questions that follow won't cover all of the definitions but only the more pertinent ones. The answers given here are the author's. Refer to Article 100 of the NEC for the official definitions. Some definitions appear in the *Code* elsewhere than Article 100 (see NEC, *Index*).

1-1   What is "accessible" as applied to wiring methods?
   Wiring that is readily available to inspection, repair, removal, etc., without disturbing the building structure or finish. Not permanently enclosed by the structure or finish of buildings.

1-2   What is "accessible" as applied to equipment?
   Equipment that may be readily reached without climbing over obstacles or that is not in locked or other hard to reach areas. For example, panelboards in kitchen cabinets that are mounted in or on the walls above washers and dryers, or in closets or bathrooms, are not accessible. Service-entrance equipment that can be reached only by going into a closet, behind a stairway, or around some other obstacle would not be considered accessible.

1-3   What does "ampacity" mean?
   The amount of flowing current (in amperes) that a conductor can carry continuously for specific use conditions and not exceed the temperature rating of the conductor (see NEC *Section 310.10*).

1-4   What is a building?
   A building may be a structure that stands by itself, or one that is separated from another by a fire wall.

1-5   What does "dead front" mean?
   No live (energized) parts are exposed where a person operates electrical equipment.

1-6   What does "approved" mean?
   Any appliance, wiring material, or other electrical equipment that is acceptable to the enforcing authorities. Underwriters Laboratory is (by far) the most acceptable authority to inspectors.

A note of caution: Don't be fooled by a UL label on the motor or cord of an appliance if there is no such label on the entire article (see NEC, *Section 110.2*). The entire appliance must be approved, not just the cord or motor.

1-7   What does "identified" mean as applied to equipment?

It means that the equipment will be suitable for the particular use or environment, that it has been evaluated by a qualified electrical testing organization, and that it features a product listing or label indicating its suitability.

1-8   What is a branch circuit?

A branch circuit is the portion of a wiring system that extends beyond the last overcurrent-protective device. In interpreting this, you must not consider the thermal cutout or the motor overload protection as the beginning of the branch circuit. The branch circuit actually begins at the final fusing or circuit-breaker point where the circuit breaks off to supply the motor.

1-9   What is a small appliance branch circuit?

This is the circuit supplying one or more outlets connecting appliances only. There is no permanently connected lighting on this circuit, except the lighting that may be built into an appliance. This term is most often used in connection with *Sections 210.11(C)(1)* and *210.52(B)* of the NEC, which refer to outlets for small appliance loads in kitchens, laundries, pantries, and dining and breakfast rooms of dwellings.

1-10   What is a general purpose circuit?

This is a branch circuit to which lighting and/or appliances may be connected. It differs from the small appliance branch circuit in question 1-9, where lighting cannot be connected.

1-11   What is a multiwire branch circuit?

A multiwire branch circuit has two or more ungrounded conductors with a potential difference between them and also has a grounded (neutral) conductor with an equal potential difference between it and each of the other wires. Examples are a three-wire 120/240-volt system or a 120/208-volt wye system, using two- or three-phase conductors and a grounded conductor. However, in either case, the "hot" wires must not be tied to one phase but must be connected to different phases to make the system a multiwire circuit (see NEC, *Section 210.4*).

1-12 What is a circuit breaker?

A device that is designed not only to open and close a circuit nonautomatically but also to open the circuit automatically at a predetermined current-overload value. The circuit breaker may be thermally or magnetically operated. However, ambient temperatures affect the operation of the thermally operated type, so that the trip value of the current is not as stable as with the magnetic type.

1-13 What is a current-carrying conductor?

A conductor that is expected to carry current under normal operating conditions.

1-14 What is a noncurrent-carrying conductor?

One that carries current only in the event of a malfunction of equipment or wiring. An equipment grounding conductor is a good example. It is employed for protection and is quite a necessary part of the wiring system, but it is not used to carry current except in the case of faulty operation, where it aids in tripping the overcurrent-protective device.

1-15 What is a pressure connector (solderless)?

A device that establishes a good electrical connection between two or more conductors by some means of mechanical pressure. A pressure connector is used in place of soldering connections and is required to be of an approved type.

1-16 What is meant by "demand factor"?

This is the ratio between the maximum demand on a system or part of a system and the total connected load on the same system or part of the system.

1-17 What is meant by "dust-tight"?

The capacity to keep dust out of an enclosure (e.g., a case or cabinet) so that dust cannot interfere with the normal operation of equipment. This is discussed further in connection with *Articles 500* and *502* of the NEC, which cover hazardous (classified) locations.

1-18 What is meant by "explosionproof apparatus"?

Apparatus enclosed in a case that is able to sustain an explosion that may occur within the case and is also able to prevent ignition of specified gases or vapors surrounding the enclosure caused by sparks, flashes, or explosion of the gases or vapors within. It must also operate at a temperature that won't ignite any inflammable atmosphere or residue surrounding it. If an explosion does occur within the equipment, the gases are allowed to escape either by a ground joint or by threads, and the escaping gases are thereby cooled to a temperature low enough to inhibit the ignition of any external gases.

1-19 What is a feeder?

The circuit conductors between the service equipment, or the source of a separately derived system, and the final branch-circuit overcurrent device or devices. Generally, feeders are comparatively large in size and supply a feeder panel, which is composed of a number of branch-circuit overcurrent devices (see *Article 215* of the NEC.)

1-20 What is a fitting?

A mechanical device, such as a locknut or bushing, that is intended primarily for a mechanical, rather than an electrical, function.

1-21 What is meant by a "ground"?

An electrical connection, either accidental or intentional, that exists between an electrical circuit or equipment and the earth, or some other conducting body that serves in place of the earth.

1-22 What does "grounded" mean?

Connected to the earth or to some other conducting body that serves in place of the earth.

1-23 What is a grounded conductor?

A system or circuit conductor that has been intentionally grounded.

1-24 What is a grounding conductor?

A conductor that is used to connect equipment, devices, or wiring systems with grounding electrodes.

1-25 What is a grounding conductor (equipment)?

The conductor used to connect noncurrent-carrying metal parts of equipment, raceways, and other enclosures to the system grounding conductor at the service and/or the grounding electrode conductor.

1-26 What is a grounding electrode conductor?
A conductor used to connect the grounding electrode to the equipment grounding conductor and/or to the grounded conductor of the circuit at the service.

1-27 What is a dwelling unit?
A dwelling unit includes one or more rooms used by one or more persons, with space for sleeping, eating, living, and a permanent provision for cooking and sanitation.

1-28 What is an outlet?
A point in the wiring system at which current is taken to supply some equipment.

1-29 What is meant by "rain-tight"?
Capable of withstanding a beating rain without allowing water to enter.

1-30 What is a receptacle?
A receptacle is a contact device installed at the outlet for the connection of a single attachment plug. A single receptacle is a single device with no other contact device on the same yoke. A multiple receptacle is a single device containing two or more receptacles.

1-31 What does "rainproof" mean?
So constructed, protected, or treated as to prevent rain from interfering with the successful operation of the apparatus.

Note
Pay particular attention to the following questions; they involve **services** and are probably among the most misused of any definitions in the NEC.

1-32 What is meant by the term "service"?
The conductors and equipment for delivering electrical energy from the secondary distribution system—the street main, the distribution feeder, or the transformer—to the wiring system on the premises. This includes the service-entrance equipment and the grounding electrode.

1-33 What are service conductors?
The portion of the supply conductors that extend from the street main, duct, or transformers to the service-entrance

equipment of the premises supplied. For overhead conductors, this includes the conductors from the last line pole (not including the service pole) to the service equipment.

1-34 What is a service cable?
A service conductor manufactured in the form of a cable and normally referred to as SE cable, or USE cable (see the NEC, *Article 338*).

1-35 What is meant by "service drop"?
The overhead conductors from the last pole or other aerial support to and including the splices, if any, connecting to the service-entrance conductors at the building or other structure. If there is a service pole with a meter on it, such as a farm service pole, the service drop does not stop at the service pole; all wires extending from this pole to a building or buildings are service drops, as well as the conductors from the last line pole to the service pole (see the NEC, *Article 100* and *Article 230, II*).

1-36 What are service-entrance conductors (overhead system)?
That portion of the service conductors between the terminals of service equipment and a point outside the building, clear of building walls, where they are joined by a splice or tap to the service drop, street main, or other source of supply.

1-37 What are service-entrance conductors (underground system)?
The service conductors between the terminals of the service equipment and the point of connection to the service lateral. Where service equipment is located outside the building walls, there may be no service-entrance conductors, or they may be entirely outside the building.

1-38 What are sets of service-entrance conductors?
Sets of service-entrance conductors are taps that run from main service conductors to service equipment.

1-39 What is meant by "service equipment"?
The necessary equipment, usually consisting of circuit breakers or switches and fuses and their accessories, located near the point of where supply conductors enter a building, structure, or an otherwise defined area, and intended to constitute the main control and means to cut off the supply.

1-40 What is meant by "service lateral"?

The underground service conductors between the street main, including any risers at the pole or other structure, or from transformers, and the first point of connection to the service-entrance conductors in a terminal box. The point of connection is considered to be the point where the service conductors enter the building.

1-41 What is meant by "service raceway"?
The rigid metal conduit, electrical metallic tubing, or other raceway that encloses service-entrance conductors.

1-42 What is meant by "special permission"?
The written consent of the authority enforcing the NEC. Under most circumstances, this is a local electrical inspector.

1-43 What is meant by a "general-use switch"?
A device intended for use as a switch in general distribution and branch circuits. It is rated in amperes and is capable of interrupting its rated current at its rated voltage.

1-44 What is meant by a "T-rated switch"?
An AC general-use snap switch that can be used (a) on resistive and inductive loads that don't exceed the ampere rating at the voltage involved, (b) on tungsten-filament lighting loads that don't exceed the ampere rating at 120 volts, and (c) on motor loads that don't exceed 80% of their ampere rating at the rated voltage.

1-45 What is meant by an "isolating switch"?
A switch that is intended to isolate an electric circuit from its source of power. It has no interrupting rating and is intended to be operated only after the circuit has been opened by some other means.

1-46 What is meant by a "motor-circuit switch"?
A switch, rated in horsepower, that is capable of interrupting the maximum operating overload current of a motor of the same horsepower rating as the switch, at the rated voltage.

1-47 What is meant by "watertight"?
So constructed that moisture won't enter the enclosing case.

1-48 What is meant by "weatherproof"?
So constructed or protected that exposure to the weather won't interfere with successful operation. Raintight or watertight may

fulfill the requirements for "weatherproof." However, weather conditions vary, and consideration should be given to conditions resulting from snow, ice, dust, and temperature extremes.

1-49 What is meant by the "voltage" of a circuit?

This is the greatest effective difference of potential (root-mean-square difference of potential) that exists between any two conductors of a circuit. On various systems, such as 3-phase 4-wire, single-phase 3-wire, and 3-wire direct current, there may be various circuits with a number of voltages.

Chapter 2

Ohm's Law and Other Electrical Formulas

When a current flows in an electric circuit, the magnitude of the current is determined by dividing the electromotive force (volts, designated by the letter E) in the circuit by the resistance (ohms, designated by the letter R) of the circuit. The resistance is dependent on the material, cross section, and length of the conductor. The current is measured in amperes and is designated by the letter I. The relationship between an electric current (I), the electromotive force (E), and the resistance (R) is expressed by Ohm's law. The following equations take into consideration only pure resistance (i.e., not inductance or capacitance); therefore, they are customarily known as the dc formulas for Ohm's law. However, in most calculations for dc circuits, which is the ordinary wiring application, these formulas are quite practical to use. Later in this book, other forms of Ohm's law that deal with inductive and capacitive reactance will be discussed.

2-1 What are the three equations for Ohm's law and what do the letters in the formulas mean?

$$I = \frac{E}{R} \quad R = \frac{E}{I} \quad E = IR$$

where I is the current flow in amperes, E is the electromotive force in volts, and R is the resistance in ohms.

2-2 A direct-current circuit has a resistance of 5 ohms. If a voltmeter connected across the terminals of the circuit reads 10 volts, how much current is flowing?
From Ohm's law, the current is:

$$I = \frac{E}{R} = \frac{10}{5} = 2 \text{ amperes}$$

2-3 If the resistance of a circuit is 25 ohms, what voltage is necessary for a current flow of 4 amperes?

From Ohm's law:

$$E = I \times R = 4 \times 25 = 100 \text{ volts}$$

2-4 If the potential across a circuit is 40 volts and the current is 5 amperes, what is the resistance?

From Ohm's law:

$$R = \frac{E}{I} = \frac{40}{5} = 8 \text{ ohms}$$

Series Circuits

A series circuit may be defined as one in which the resistive elements are connected in a continuous run (i.e., connected end to end) as shown in Figure 2-1. It is evident that since the circuit has only one pathway (no branches), the amount of current flowing must be the same in all parts of the circuit. Therefore, the current flowing through each resistance is also equal. The total potential across the entire circuit equals the sum of potential drops across each individual resistance, or:

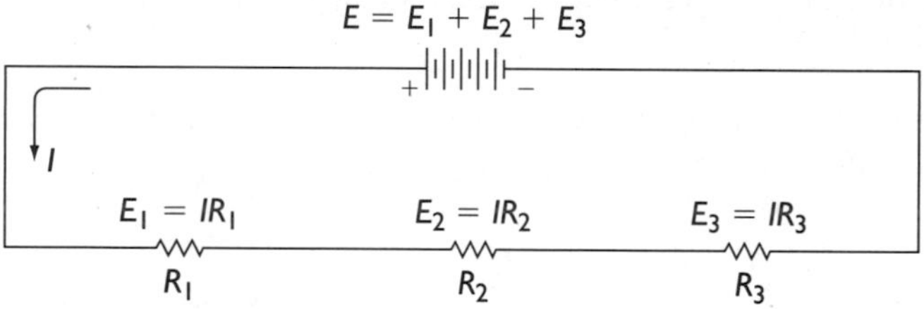

$$E = E_1 + E_2 + E_3$$

Figure 2-1 Resistances in series.

$$E_1 = IR_1$$
$$E_2 = IR_2$$
$$E_3 = IR_3$$
$$E = E_1 + E_2 + E_3$$

and

$$R = R_1 + R_2 + R_3$$

The equation for the total potential of the circuit is:

$$E = IR_1 + IR_2 + IR_3 = I(R_1 + R_2 + R_3)$$

and

$$I = \frac{E}{R_1 + R_2 + R_3} = \frac{E}{R}$$

2-5 If the individual resistances shown in Figure 2-1 are 5, 10, and 15 ohms, respectively, what potential must the battery supply to force a current of 0.5 ampere through the circuit?

The total resistance is:

$$R = 5 + 10 + 15 = 30 \text{ ohms}$$

Hence, the total voltage is:

$$E = 0.5 \times 30 = 15 \text{ ohms}$$

As a check, calculate the individual voltage drop across each part:

$$E_1 = 0.5 \times 5 = 2.5 \text{ volts}$$
$$E_2 = 0.5 \times 10 = 5.0 \text{ volts}$$
$$E_3 = 0.5 \times 15 = 7.5 \text{ volts}$$

and,

$$E = E_1 + E_2 + E_3 = 2.5 + 5.0 + 7.5 = 15 \text{ volts}$$

2-6 In order to determine the voltage of a dc source, three resistance units of 10, 15, and 30 ohms are connected in series with this source. If the current through the circuit is 2 amperes, what is the potential of the source?

$$E = I(R_1 + R_2 + R_3)$$
$$= 2(10 + 15 + 30) = 2 \times 55$$
$$= 110 \text{ volts}$$

Parallel Circuits

In a parallel, or divided, circuit such as that shown in Figure 2-2, the same voltage appears across each resistance in the group; the current flowing through each resistance is inversely proportional to the value of the resistance. The sum of all the currents, however, is equal to the total current leaving the battery. Thus:

$$E = I_1 R_1 = I_2 R_2 = I_3 R_3$$

and,

$$I = I_1 + I_2 + I_3$$

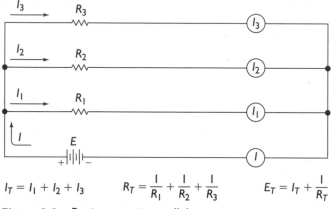

$$I_T = I_1 + I_2 + I_3 \qquad R_T = \frac{1}{R_1} + \frac{1}{R_2} + \frac{1}{R_3} \qquad E_T = I_T + \frac{1}{R_T}$$

Figure 2-2 Resistances in parallel.

When Ohm's law is applied to the individual resistances, the following equations are obtained:

$$I_1 = \frac{E}{R_1} \quad I_2 = \frac{E}{R_2} \quad I_3 = \frac{E}{R_3}$$

Hence,

$$I_1 = \frac{E}{R_1} + \frac{E}{R_2} + \frac{E}{R_3}$$

or

$$I = E\left(\frac{1}{R_1} + \frac{1}{R_2} + \frac{1}{R_3}\right)$$

and since $I = E/R$, the equivalent resistance of the several resistances connected in parallel is:

$$\frac{1}{R} = \frac{1}{R_1} + \frac{1}{R_2} + \frac{1}{R_3}$$

You've found, then, that any number of resistances in parallel can be replaced by an equivalent resistance whose value is equal to the reciprocal of the sum of the reciprocals of the individual resistances. You will find that the sum of resistances in parallel will always be smaller than the value of the smallest resistor in the group.

The value 1/R, or the reciprocal of the value of the resistance, is expressed as the conductance of the circuit; its unit is mho (ohm spelled backward) and is usually expressed by g or G.

Where there are only two resistances connected in parallel,

$$R = \frac{R_1 \times R_2}{R_1 + R_2} \text{ ohms}$$

Where there are any number of equal resistances connected in parallel, you may divide the value of one resistance by the number of equal resistances.

2-7 A resistance of 2 ohms is connected in series with a group of three resistances in parallel, which are 4, 5, and 20 ohms, respectively. What is the equivalent resistance of the circuit?

The equivalent resistance of the parallel network is:

$$\frac{1}{R} = \frac{1}{4} + \frac{1}{5} + \frac{1}{20} = 0.25 + 0.20 + 0.05 = 0.50$$

$$R = \frac{1}{0.50} = 2 \text{ ohms}$$

The circuit is now reduced to two series resistors of 2 ohms each, as shown in Figure 2-3. The equivalent resistance of the circuit is 2 + 2, or 4 ohms.

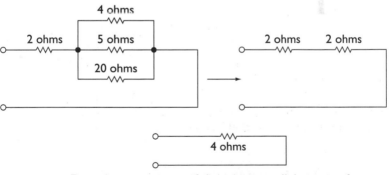

Figure 2-3 Equivalent resistance of the series-parallel circuit of question 2-7.

2-8 Two parallel resistors of 2 and 6 ohms are connected in series with a group of three parallel resistors of 1, 3, and 6 ohms, respectively. If the two parallel-resistance groups are connected in series by means of a 1.5-ohm resistor, what is the equivalent resistance of the system?

Replace the 2- and 6-ohm resistors by a resistance of R_1, where:

$$R_1 = \frac{2 \times 6}{2 + 6} = \frac{12}{8} = 1.5 \text{ ohms}$$

Replace the group of three resistors by R_2, where:

$$\frac{1}{R} = \frac{1}{1} + \frac{1}{3} + \frac{1}{6} = 1.0 + 0.33 + 0.17 = 1.50$$

$$R = \frac{1}{1.50} = 0.67 \text{ ohms}$$

The circuit is now reduced to three series resistances, the values of which are 1.5, 1.5 and 0.67 ohms, as shown in Figure 2-4. Their combined values are:

$$R_T = R + R_1 + R_2$$
$$R_T = 1.5 + 1.5 + 0.67 = 3.67 \text{ ohms}$$

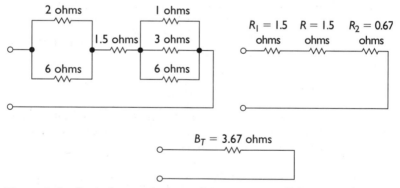

Figure 2-4 Equivalent resistance of the series-parallel circuit of question 2-8.

Units of Area and Resistance

The circular mil is the unit of cross section used in the American wire gauge (AWG) or the B&S wire gauge systems (see NEC, *Chapter 9, Table 8*). The term *mil* means one thousandth of an inch (0.001 inch). A circular mil is the area of the cross-sectional surface of a cylindrical wire with a diameter of 1 mil (0.001 inch).

The circular mil area of any solid cylindrical wire is equal to the wire's diameter (expressed in mils) squared. For example, the area in circular mils (written CM or cir. mils) of a wire having a diameter of ⅜ inch (0.375) equals $375 \times 375 = 375^2 = 140,625$ CM. The diameter in mils of a solid circular wire is equal to the square root of its circular mil area. Assuming that a conductor has an area of 500,000 CM, its diameter in mils is the square root of 500,000, which is equal to 707 mils, or 0.707 inch (approximately). The area in square inches of a wire whose diameter is 1 mil is:

$$\frac{\pi}{4} = 0.7845 \times 0.001^2 = 0.0000007854 \text{ sq. in.}$$

The square mil is the area of a square whose sides are each 1 mil (0.001 inch). Hence, the area of a square mil is 0.001^2, or 0.000001 square inch. With reference to the previous definitions of the circular mil and the square mil, it is obvious that in order to convert a unit of circular area into its equivalent area in square mils, the circular mil must be multiplied by $\pi/4$, or 0.7854, which is the same as dividing by 1.273. Conversely, to convert a unit area of square mils into its equivalent area in circular mils, the square mil should be divided by $\pi/4$, or 0.7854, which is the same as multiplying by 1.273.

The above relations may be written as follows:

$$\text{Square mils} = \text{circular mils} \times 0.7854 - \frac{\text{circular mils}}{1.273}$$

$$\text{Circular mils} = \frac{\text{square mils}}{0.7854} = \text{square mils} \times 1.273$$

Thus, any circular conductor may be easily converted into a rectangular conductor (a bus bar, for example) containing an equivalent area of current-carrying capacity.

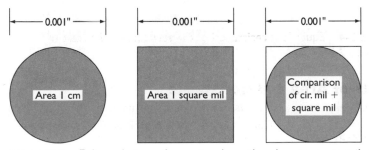

Figure 2-5   Enlarged view of one circular mil and one square mil, with a comparison of the two.

2-9 If a No. 10 wire (B&S or AWG) has a diameter of 101.9 mils, what is its circular mil area?

$$\text{Area} = 101.9^2 = 10,383.6 \text{ CM}$$

Referring to Table 2-1, you will find this to be approximately true. If you concentrate on remembering the CM area of No. 10 wire as 10,380 CM, you will find this invaluable in arriving at the CM area of any size wire without the use of a table. For practical purposes, if you do not have a wire table such as Table 2-1 readily available, you may find the circular mil area of any wire size by juggling back and forth as follows, and the answer will be close enough for all practical purposes.

> Every three sizes removed from No. 10 doubles or halves the area in CM.

> Every ten sizes removed from No. 10 is 1/10 or 10 times the area in CM.

Example: Ten sizes smaller than No. 10 is No. 0 wire. No. 10 wire has a circular mil area of 10,380 CM; therefore, No. 0 would be 103,800 CM. Table 2-1 shows an area of 105,600 CM for No. 0 wire, or about a 1.7% error.

Example: Ten sizes smaller than No. 10 is No. 20 wire. Number 20 wire should then have a circular mil area of 1038.1 CM.

You can see from the examples above that for quick figuring, the percentage of error is quite small. However, you should always have a wire table nearby. (You can find a pretty good one in *Chapter 9, Table 8* of the NEC.)

2-10 Number 000 wire (AWG) has an area of 167,800 CM. If it were solid copper (not made of strands), what would its diameter be?
The diameter is:

$$\sqrt{167,800} = 409.6 \text{ mils, or } 0.4096 \text{ inch}$$

2-11 A certain switchboard arrangement necessitates conversion from a circular conductor to a rectangular bus bar having an equivalent area. If the diameter of the solid conductor measures 0.846 inch, calculate (a) the width of an equivalent bus bar, if the thickness of the bus bar is ¼ inch; (b) the bus bar area in circular

mils; (c) the current-carrying capacity, if 1 square inch of copper carries 1000 amperes.

The area, in square inches, of the circular conductor is:

$$A = 0.7854 \times D^2 = 0.7854 \times 0.846^2$$
$$= 0.562 \text{ square inch}$$

The area, in square inches, of the bus bar is:

$$A' = 0.25 \times W$$

where W is the width of the bus bar.
Since,

$$A' = A\, 0.562 = 0.25\,W$$

(a) $W = 0.562/025 = 2.248$ inches

(b) The area of the conductor $= 846^2 = 715,716$ CM

(c) The current-carrying capacity of the conductor $= 0.562 \times 1000 = 562$ amps.

2-12 A certain 115-volt, 100-horsepower dc motor has an efficiency of 90% and requires a starting current that is 150% of the full-load current. Determine (a) the size of fuse needed, in amps; (b) the copper requirements of the switch. Assume that 1 square inch of copper carries 1000 amperes.

Motor current:

$$I_M = \frac{\text{hp} \times 746}{E \times \text{efficiency}} = \frac{100 \times 746}{115 \times 0.9} = 721 \text{ amperes}$$

(a) The amperage of the fuses is, therefore, $721 \times 1.5 = 1081.5$, or 1200 amperes, which is the rating of the closest manufactured fuse.

(b) Since the motor required a current of 721 amperes, the copper of each switch blade must be $^{721}/_{1000}$, or 0.721 square inch. Therefore, if $\frac{3}{8}$-inch bus copper is used, its width must be:

$$W = \frac{0.721}{0.375} = 1.92 \text{ inches}$$

The mil-foot is a unit of cross-sectional area of a cylindrical conductor that is one foot in length and one mil in diameter. The resistance of such a unit of copper has been found experimentally to be 10.37 ohms at 20°C; this is normally thought of as 10.4 ohms (Figure 2-6).

A mil-foot of copper at 20°C offers 10.4 ohms resistance; at 30°C, it is 11.2 ohms; at 40°C, it is 11.6 ohms; at 50°C, it is 11.8 ohms; at 60°C, it is 12.3 ohms; and at 70°C, it is 12.7 ohms. Thus, in voltage-drop calculations, 12 is generally used as the constant K, unless otherwise specified, to allow for higher temperatures and to afford some factor of safety.

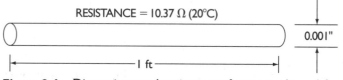

Figure 2-6 Dimensions and resistance of one circular mil-foot of copper.

The resistance of a wire is directly proportional to its length and inversely proportional to its cross-sectional area. Therefore, if the resistance given in ohms of a mil-foot of wire is multiplied by the total length in feet (remember that there are practically always two wires, so if the distance is given in feet, multiply it by two to get the total resistance of both wires; this is a common error when working examination problems) and divided by its cross-sectional area in circular mils, the result will be the total resistance of the wire in ohms. This is expressed as:

$$R = \frac{K \times L \times 2}{A}$$

where: R is the resistance in ohms
K is the constant (12) for copper
L is the length in feet one way
A is the area in circular mils

K for copper was given, so it will be well to give K for commercial aluminum:

$$20°C(68°F): K = 17.39 \quad 50°C (122°F): K = 19.73$$
$$30°C(86°F): K = 18.73 \quad 60°C (140°F): K = 20.56$$
$$40°C(104°F): K = 19.40 \quad 70°C (158°F): K = 21.23$$

$K = 12$ was used for copper, because in most instances it provides a safety factor in figuring voltage drop. Thus, with a correction factor of 1.672 for aluminum, as opposed to copper, it would be well to use a K factor of 20 for aluminum.

Should you have other conditions, K factors have been given for both copper and aluminum at various degrees Celsius. To change Celsius to Fahrenheit, use the following formula:

$$\text{Degrees C} \times 1.8 + 32 = \text{Degrees F}$$

Thus:

$$30°C \times 1.8 + 32 = 54 + 32 = 86°F$$

2-13 What is the resistance of a 500-foot line of No. 4 copper wire?

From *Chapter 9, Table 8* of the NEC, No. 4 wire has a cross-sectional area of 41,740 CM. Therefore,

$$R = \frac{12 \times 500 \times 2}{41,740} = \frac{12,000}{41,740} = 0.29 \text{ ohms}$$

Table 8 lists a value of 0.2480 ohm for 1000 feet of No. 4 wire at 20°C. The difference is that a K of 12 was used instead of 10.4; 20°C is 68°F, and the 12 value is for slightly over 50°C, or 122°F.

2-14 Suppose it is desired to have a copper wire of 0.5-ohm resistance whose total length is 2000 feet, or 1000 feet one way. What must its cross-sectional area be? What size wire is necessary?

$$A = K\frac{L}{R} = \frac{12 \times 2000}{0.5} = 48,000 \text{ CM}$$

According to *Chapter 9, Table 8* of the NEC, No. 4 wire has a cross-sectional area of 41,740 CM and No. 3 wire has an area of 52,620 CM, so use the larger size No. 3 wire. Also, notice that the equation did not use the factor 2 because the 2000-foot length was the total and not just the length of one wire.

2-15 If the resistance of a copper wire whose diameter is ⅛ inch is measured as 0.125 ohm, what is the length of the wire?

$$L = \frac{RA}{K} = \frac{0.125 \times 125^2}{12} = 163 \text{ feet}$$

Here again, this is the total length; one way would be 81.5 feet.

2-16 A copper line that is 5 miles in length has a diameter of 0.25 inch. Calculate: (a) the diameter of the wire in mils; (b) the area of the wire in circular mils; (c) the weight in pounds; (d) the resistance at 50°C. Assume that the weight in pounds per cubic inch is 0.321.

(a) The diameter in mils = 1000 × 0.25 = 250 mils

(b) The area in CM = 250^2 = 62,500 CM

(c) Cross-sectional area = 0.7854 × D^2 = 0.7854 × 0.25^2 = 0.0491 sq. in. Length of wire = 5280 × 5 × 2 × 12 = 633,600 in. Weight of wire = 0.0491 × 633,600 × 0.321 = 9,986.23, or about 10,000 lb.

(d)
$$R = \frac{K \times 2 \times (5 \times 5280)}{62,500}$$
$$= \frac{12 \times 2 \times 26,400}{62,500}$$
$$= 10.14 \text{ ohms}$$

Skin Effect

When alternating current flows through a conductor, an inductive effect occurs, which tends to force the current to the surface of the conductor. This produces a voltage loss and also affects the current-carrying capacity of the conductor. For open wires or wires in nonmetallic-sheathed cable, this "skin effect" is neglected until the No. 0 wire size is reached. In metallic-sheathed cables and metallic raceways, the skin effect is neglected until size No. 2 is reached. At these points, there are multiplying factors for conversion from dc resistance to ac resistance (Table 2-1). Note that there is a different factor for aluminum than for copper cables.

2-17 The dc resistance of a length of 250,000-CM copper cable in rigid metal conduit was found to be 0.05 ohm. What would its ac resistance be?

From Table 2-1, the multiplying factor is found to be 1.06; therefore, R_{AC} = 0.05 ohm × 1.06 = 0.053 ohm.

Table 2-1   Multiplying Factors for Converting dc Resistance to 60-Cycle ac Resistance

| | Multiplying Factor | | | |
| Size | For Nonmetallic-Sheathed Cables in Air or Nonmetallic Conduit | | For Metallic-Sheathed Cables or All Cables in Metallic Raceways | |
	Copper	Aluminum	Copper	Aluminum
Up to 3 AWG	1.000	1.000	1.00	1.00
2	1.000	1.000	1.01	1.00
1	1.000	1.000	1.01	1.00
0	1.001	1.000	1.02	1.00
00	1.001	1.001	1.03	1.00
000	1.002	1.001	1.04	1.01
0000	1.004	1.002	1.05	1.01
250 kcmil	1.005	1.002	1.06	1.02
300 kcmil	1.006	1.003	1.07	1.02
350 kcmil	1.009	1.004	1.08	1.03
400 kcmil	1.011	1.005	1.10	1.04
500 kcmil	1.018	1.007	1.13	1.06
600 kcmil	1.025	1.010	1.16	1.08
700 kcmil	1.034	1.013	1.19	1.11
750 kcmil	1.039	1.015	1.21	1.12
800 kcmil	1.044	1.017	1.22	1.14
1000 kcmil	1.067	1.026	1.30	1.19
1250 kcmil	1.102	1.040	1.41	1.27
1500 kcmil	1.142	1.058	1.53	1.36
1750 kcmil	1.185	1.079	1.67	1.46
2000 kcmil	1.233	1.100	1.82	1.56

Voltage-Drop Calculations

The methods for finding the resistance of wire have been discussed. Now you can use Ohm's law to find the voltage drop for circuits loads. Use the form

$$E = I \times R$$

Under *Section 215.2(A)(4)*, FPN 2 of the NEC, find the prescribed maximum allowable percent of voltage drop permitted for feeders and branch circuits. In *Section 210.19(A)*, FPN 4 of the NEC, the maximum allowable percent of voltage drop for branch circuits is given.

2-18 What is the percentage of allowable voltage drop for feeders that are used for power and heating loads?
Maximum of 3%.

2-19 What is the percentage of allowable voltage drop for feeders that are used for lighting loads?
Maximum of 3%.

2-20 What is the percentage of allowable voltage drop for combined lighting, heating, and power loads?
A maximum 3% voltage drop for feeders and 5% for feeders and branch circuits is allowable. The following formula is used for voltage drop calculations.
Since,

$$R = \frac{K \times L \times 2}{A}$$

then,

$$E_d = I \times R = \frac{K \times 2L \times I}{A}$$

where E_d is the voltage drop of the circuit
$2L$ is the total length of the wire
K is a constant (12)
I is the current, in amperes, of the circuit
A is the area, in circular mils, of the wire in the circuit

By transposing the formula above, determine the circular mil area of a wire for a specified voltage drop:

$$A = \frac{K \times 2L \times I}{E_d}$$

2-21 A certain motor draws 22 amps at 230 volts, and the feeder circuit is 150 feet in length. If No. 10 copper wire is desired, what would the voltage drop be? Would No. 10 wire be permissible to use?

$$E_d = \frac{12 \times 2 \times 150 \times 22}{10,380}$$
$$= \frac{79,200}{10,380}$$
$$= 7.63 \text{ volts}$$

However, $230 \times 0.03 = 6.90$ volts, which is the voltage drop permissible under *Section 215.2* of the NEC, so No. 10 wire would not be large enough.

2-22 In question 2-21, a voltage drop of 7.63 volts was calculated; however, the maximum permissible voltage drop is 6.9 volts. What size wire would have to be used?

According to *Chapter 9, Table 8* of the NEC, No. 10 wire has an area of 10,380 CM, and No. 9 wire has an area of 13,090 CM. Therefore, No. 9 would be the proper size, except that you cannot purchase No. 9 wire; you will have to use No. 8 wire, which has an area of 16,510 CM.

2-23 If No. 8 copper wire were used in question 2-22, what would the voltage drop be? What percentage would this drop be?

$$E_d = \frac{12 \times 2 \times 150 \times 22}{16,509} = \frac{79,200}{16,509} = 4.80 \text{ volts}$$

$$\frac{4.80}{2.30} \times 100 = 2.09\%$$

$$A = \frac{K \times 2L \times I}{E_d}$$

$$= \frac{12 \times 2 \times 150 \times 22}{6.9}$$

$$= 11,478.3 \text{ CM}$$

Formulas for Determining Alternating Current in Alternating-Current Circuits

In the formulas of Figure 2-7:

R is the resistance in ohms

X_L is the inductive reactance in ohms $= 2\pi fL$

X_C is the capacitive reactance in ohms $= 2\pi fC$

f is the frequency

L is the inductance in henrys

C is the capacity in farads

Z is the impedance in ohms

I is the current in amperes

E is the pressure in volts

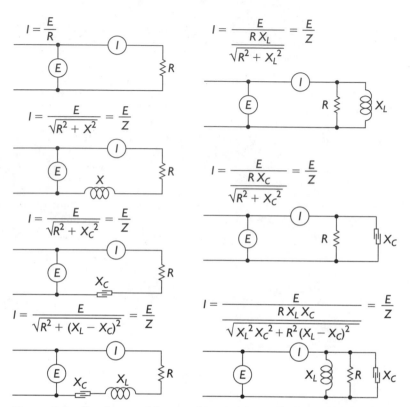

Figure 2-7 Fundamental forms of ac circuits, with the method of determining current when voltage and impedance are known.

Formulas for Combining Resistance and Reactance

In the following formulas of Figure 2-8:

> R is the resistance in ohms
> X_L is the inductive reactance in ohms $= 2\pi fL$
> X_C is the capacitive reactance in ohms $= 1/2\pi fC$
> f is the frequency
> L is the inductance in henrys
> C is the capacity in farads
> Z is the impedance in ohms
> I is the current in amperes
> E is the pressure in volts

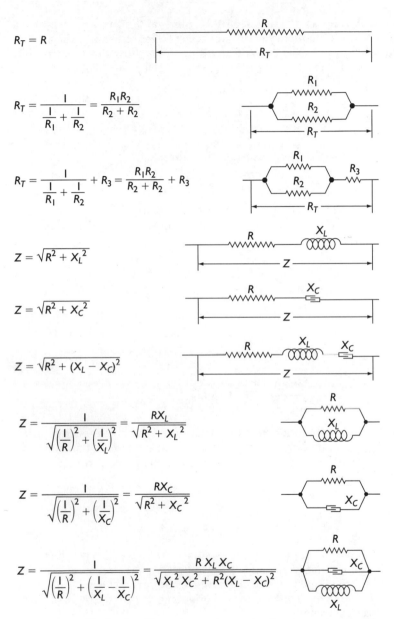

$$R_T = R$$

$$R_T = \frac{1}{\frac{1}{R_1} + \frac{1}{R_2}} = \frac{R_1 R_2}{R_2 + R_2}$$

$$R_T = \frac{1}{\frac{1}{R_1} + \frac{1}{R_2}} + R_3 = \frac{R_1 R_2}{R_2 + R_2} + R_3$$

$$Z = \sqrt{R^2 + X_L^2}$$

$$Z = \sqrt{R^2 + X_C^2}$$

$$Z = \sqrt{R^2 + (X_L - X_C)^2}$$

$$Z = \frac{1}{\sqrt{\left(\frac{1}{R}\right)^2 + \left(\frac{1}{X_L}\right)^2}} = \frac{R X_L}{\sqrt{R^2 + X_L^2}}$$

$$Z = \frac{1}{\sqrt{\left(\frac{1}{R}\right)^2 + \left(\frac{1}{X_C}\right)^2}} = \frac{R X_C}{\sqrt{R^2 + X_C^2}}$$

$$Z = \frac{1}{\sqrt{\left(\frac{1}{R}\right)^2 + \left(\frac{1}{X_L} - \frac{1}{X_C}\right)^2}} = \frac{R X_L X_C}{\sqrt{X_L^2 X_C^2 + R^2(X_L - X_C)^2}}$$

Figure 2-8 Methods of impedance determination in ac circuits when resistance, inductive reactance, and capacitive reactance (in ohms) are known.

2-24 A coil has a resistance of 10 ohms and an inductance of 0.1 henry. If the frequency of the source is 60 hertz, what is the voltage necessary to cause a current of 2 amperes to flow through the coil?

With reference to the previously derived formula,

$$E_R = IR = 2 \times 10 = 20 \text{ volts}$$
$$X_L = 2\pi fL = 2\pi \times 60 \times 0.1 = 37.7 \text{ ohms}$$
$$E_L = IX_L = 2 \times 37.7 = 75.4 \text{ volts}$$

The applied voltage must therefore be:

$$E = \sqrt{E_R^2 + E_L^2} = \sqrt{20^2 + 75.4^2} = 78 \text{ volts}$$

2-25 A coil with a negligible resistance requires 3 amperes when it is connected to a 180-volt, 60-hertz supply. What is the inductance of the coil?

$$X_L = \frac{E}{I} = \frac{180}{3} = 60 \text{ ohms}$$

and,

$$X_L = 2\pi fL$$

therefore,

$$L = \frac{60}{2\pi \times 60} = 0.159 \text{ henry}$$

2-26 An alternating current of 15 amperes with a frequency of 60 hertz is supplied to a circuit containing a resistance of 5 ohms and an inductance of 15 millihenrys. What is the applied voltage?

$$X_L = 2\pi fL = 2\pi \times 60 \times 0.015 = 5.65 \text{ ohms}$$
$$Z = \sqrt{R^2 + X_L^2} = \sqrt{5^2 + 5.65^2} = 7.54 \text{ ohms}$$
$$E = IZ = 15 \times 7.54 = 113.1 \text{ volts}$$

2-27 A coil contains 5 ohms resistance and 0.04 henry inductance. The voltage and frequency of the source are 100 volts at 60 cycles. Find (a) the impedance of the coil; (b) the current through the coil; (c) the voltage drop across the inductance; (d) the voltage drop across the resistance.

$$X_L = 2\pi fL = 2\pi \times 60 \times 0.04 = 15 \text{ ohms}$$

(a) $Z = \sqrt{5^2 + 15^2} = \sqrt{250} = 15.8$ ohms

(b) $I = \dfrac{E}{Z} = \dfrac{100}{\sqrt{5^2 + 15^2}} = 6.3$ amperes

(c) $E_L = IX_L = 6.3 \times 15 = 94.5$ volts

(d) $E_R = IR = 6.3 \times 5 = 31.5$ volts

2-28 An alternating-current contains 10 ohms resistance in series with a capacitance of 40 microfarads. The voltage and frequency of the source are 120 volts at 60 cycles. Find (a) the current in the circuit; (b) the voltage drop across the resistance; (c) the voltage drop across the capacitance; (d) the power factor; (e) the power loss.

$$X_C = \frac{1}{2\pi fC} = \frac{1}{2\pi \times 60 \times 0.00004} = 66.67 \text{ ohms}$$
$$Z = \sqrt{10^2 + 66.7^2} = 67 \text{ ohms}$$

(a) $I = \dfrac{E}{Z} = \dfrac{120}{67} = 1.8$ amperes

(b) $E_R = IR = 1.8 \times 10 = 18$ volts

(c) $E_C = IX_C = 1.8 \times 66.3 \times 119.3$ volts

(d) $\cos \phi = \dfrac{R}{Z} = \dfrac{10}{67} = 0.15,$ or 15%

(e) $P = E \times I \times \cos \phi$
$$= 120 \times 1.8 \times 0.15$$
$$= 32.4 \text{ watts}$$

2-29 A coil of 3 ohms resistance and 20 millihenrys inductance is connected in series with a capacitance of 400 microfarads. If the voltage and frequency are 120 volts at 60 hertz, find (a) the impedance of the circuit; (b) the current in the circuit; (c) the power loss; (d) the power factor.

$$X_L = 2\pi fL = 2\pi \times 60 \times 0.020 = 7.54 \text{ ohms}$$
$$X_C = \frac{1}{2\pi \times 60 \times 0.0004} = 6.67 \text{ ohms}$$

(a) $Z = \sqrt{R^2 + (X_L - X_C)^2}$
$$= \sqrt{3^2 + (7.54 - 6.67)^2}$$
$$= 3.12 \text{ ohms}$$

(b) $I = \dfrac{E}{Z} = \dfrac{120}{3.12} = 38.5$ amperes

(c) $P = I^2R = 38.5^2 \times 3 = 4,447$ watts

(d) $\cos\phi = \dfrac{R}{Z} = \dfrac{3}{3.12} = 0.961$, or 96% (approx.)

As a check for the power loss in part (c), use the information obtained in part (d), the power factor. Then:

$$P = EI\cos\phi = 120 \times 38.5 \times 0.961 = 4,440 \text{ watts}$$

2-30 A certain series circuit has a resistance of 10 ohms, a capacitance of 0.0003 farad, and an inductance of 0.03 henry. If a 60-hertz, 230-volt Electromotive force (emf) is applied to this circuit, find (a) the current through the circuit; (b) the power factor; (c) the power consumption.

$$X_L = 2\pi fL = 2\pi \times 60 \times 0.03 = 11.30 \text{ ohms}$$

$$X_C = \frac{1}{2\pi fC} = \frac{1}{2\pi \times 60 \times 0.0003} = 8.85 \text{ ohms}$$

$$Z = \sqrt{R^2 + (X_L - X_C)^2}$$
$$= \sqrt{10^2 + (11.30 - 8.85)^2}$$
$$= 10.3 \text{ ohms}$$

(a) $I = \dfrac{E}{Z} = \dfrac{230}{10.3} = 22.3$ amperes

(b) $\cos\phi = \dfrac{10}{10.2} = 0.97$, or 97%

(c) $P = I^2R = 22.3^2 \times 10 = 4.97$ kilowatts

2-31 The circuit in Figure 2-9 contains a resistance of 30 ohms and a capacitance of 125 microfarads. If an alternating current of 8 amperes at a frequency of 60 hertz is flowing in the circuit, find (a) the voltage drop across the resistance; (b) the voltage drop across the capacitance; (c) the voltage applied across the circuit.

$$X_C = \frac{1}{2\pi fC} = \frac{1}{2\pi \times 60 \times 0.000125} = 21.3 \text{ ohms}$$

$$Z = \sqrt{R^2 + X_C^2} = \sqrt{30^2 + 21.3^2} = 36.8 \text{ ohms}$$

(a) $E_R = IR = 8 \times 30 = 240$ volts
(b) $E_C = IX_C = 8 \times 21.3 = 170$ volts
(c) $E = IZ = 8 \times 36.8 = 294$ volts

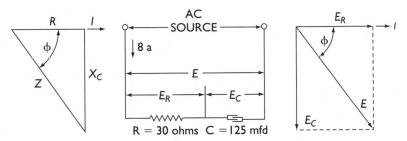

Figure 2-9 Resistance and capacitance in series, with appropriate vector diagrams.

2-32 A resistance of 15 ohms is connected in series with a capacitance of 50 microfarads. If the voltage of the source is 120 volts at 60 hertz, find (a) the amount of current in the circuit; (b) the voltage drop across the resistance; (c) the voltage drop across the capacitance; (d) the angular difference between the current and the applied voltage; (e) the power loss in the circuit.

$$X_C = \frac{1}{2\pi fC} = \frac{1}{2\pi \times 60 \times 0.00005} = 53.0 \text{ ohms}$$

$$Z = \sqrt{R^2 + X_C^2} = \sqrt{15^2 + 53.0^2} = 55.1 \text{ ohms}$$

(a) $I = \dfrac{E}{Z} = \dfrac{120}{55.1} = 2.18$ amperes

(b) $E_R = IR = 2.18 \times 15 = 32.7$ volts

(c) $E_C = IX_C = 2.18 \times 53.0 = 115.5$ volts

(d) $\cos \phi = \dfrac{E_R}{E} = \dfrac{32.7}{120} = 0.272, \quad \text{and} \quad \phi = 74.2°$

(e) $P = I^2R = 2.18^2 \times 15 = 71$ watts

2-33 A resistance of 20 ohms and a capacitance of 100 microfarads are connected in series across a 200-volt, 50-hertz ac supply. Find (a) the current in the circuit; (b) the potential drop across the resistance; (c) the potential drop across the capacitance; (d) the phase difference between the current and the applied voltage; (e) the power consumed; (f) the power factor.

$$X_C = \frac{1}{2\pi f C} = \frac{1}{2\pi \times 50 \times 0.0001} = 31.8 \text{ ohms}$$

$$Z = \sqrt{R^2 + X_C^2} = \sqrt{20^2 + 31.8^2} = 37.6 \text{ ohms}$$

(a) $I = \dfrac{E}{Z} = \dfrac{200}{37.6} = 5.32$ amperes

(b) $E = IR = 5.32 \times 20 = 106.4$ volts

(c) $E_C = IX_C = 5.32 \times 31.8 = 169.2$ volts

(d) $\cos \phi = \dfrac{R}{Z} = \dfrac{20}{37.6} = 0.532,$ and $\phi = 57.9°$

(e) $P = I^2R = 5.32^2 \times 20 = 566$ watts

(f) $\cos \phi = 0.532,$ or 53.2%

2-34 What is the total capacitance of four parallel capacitors rated 10, 15, 25, and 30 microfarads, respectively?

$$\begin{aligned} C &= C_1 + C_2 + C_3 + C_4 \\ &= 10 + 15 + 25 + 30 \\ &= 80 \text{ microfarads} \end{aligned}$$

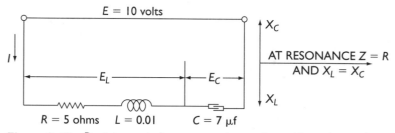

Figure 2-10   Resistance, inductance, and capacitance in series, with a vector diagram illustrating their relationship to each other.

2-35 A certain coil (shown in Figure 2-10) with a resistance of 5 ohms and an inductance of 0.01 henry is connected in series with a capacitor across a 10-volt supply, which has a frequency of 800 cycles per second. Find (a) the capacitance that will produce resonance; (b) the corresponding value of the current; (c) the potential drop across the coil; (d) the potential drop across the capacitor; (e) the power factor of the circuit; (f) the power consumption.

The inductive reactance of the coil is:

$$X_L = 2\pi \times 800 \times 0.01 = 50.24 \text{ ohms}$$

Therefore,

$$X_C = 50.24 = \frac{10^6}{2\pi \times 800 \times C}$$

$$C = \frac{10^4}{50.24^2}$$

$$= 3.96 \text{ microfarads}$$

(a) Since resonance occurs when $X_L = X_C$, X_C must also be equal to 50.24 ohms.

(b) At resonance, the current is:

$$I = \frac{E}{R} = \frac{10}{5} = 2 \text{ amperes}$$

(c) The potential drop across the coil is:

$$E_L = IX_L = 2 \times 50.24 = 100.5 \text{ volts}$$

(d) The potential drop across the capacitor is:

$$E_C = IX_C = 2 \times 50.24 = 100.5 \text{ volts}$$

(e) The power factor is:

$$\cos\phi = \frac{R}{Z}$$

but since resonance $Z = R$,

$$\cos\phi = \frac{R}{R} = 1, \text{ and } \phi = 0°$$

(f) The power consumed is:

$$P = I^2R = 4 \times 5 = 20 \text{ watts}$$

2-36 The field winding of a shunt motor has a resistance of 110 ohms, and the emf applied to it is 220 volts. What is the amount of power expended in the field excitation?

The current through the field is:

$$I_f = \frac{E_t}{R_f} = \frac{220}{110} = 2 \text{ amperes}$$

The power expended $= E_t I_f = 220 \times 2 = 440$ watts. The same results can also be obtained directly by using the following equation:

$$P_f = \frac{E_f^2}{R_f} = \frac{220^2}{110} = 440 \text{ watts}$$

2-37 A shunt motor whose armature resistance is 0.2 ohm and whose terminal voltage is 220 volts requires an armature current of 50 amperes and runs at 1500 rpm when the field is fully excited. If the strength of the field is decreased and the amount of armature current is increased, both by 50%, at what speed will the motor run?

The expression for the counter-emf of the motor is:

$$E_a = E_t = I_a R_a$$

and,

$$E_{a^1} = 220 - (50 \times 0.2) = 210 \text{ volts}$$

Similarly,

$$E_{a^2} = 220 - (75 \times 0.2) = 205 \text{ volts}$$

also,

$$E_a = NfK$$

and,

$$\frac{E_{a^1}}{E_{a^2}} = \frac{N_1 \phi_1 K_1}{N_2 \phi_2 K_2}$$

Since the field is decreased by 50%,

$$\phi_1 = 1.5\phi_2, \text{ and } Z_1 = Z_2$$

it follows that:

$$\frac{210}{205} = \frac{1500 \times 1.5}{N_2}$$

$$N_2 = \frac{1500 \times 205 \times 1.5}{210} = 2196 \text{ rpm}$$

2-38 A 7.5-hp, 220-volt interpole motor has armature and shunt-field resistances of 0.5 ohm and 200 ohms, respectively. The current input at 1800 rpm under no-load conditions is 3.5 amperes. What are the current and the electromagnetic torque for a speed of 1700 rpm?

Under no-load conditions (at 1800 rpm),

$$I_a = I_L - I_f = 3.5 - \frac{220}{200} = 2.4 \text{ amperes}$$

$$\phi K_{NL} = \frac{E_t - (I_a R_a)}{N} = \frac{220 - (2.4 \times 0.5)}{1800} = 0.1216$$

$$\phi K_{NL} = \phi K_{FL}$$

At 1700 rpm,

$$I_a = \frac{E_t - (N\phi K)}{R_a}$$

$$= \frac{220 - (1700 \times 0.1216)}{0.5}$$

$$= 26.6 \text{ amperes}$$

$$I_L = I_a + I_f = 26.6 + 1.1 = 27.7 \text{ amperes}$$

$$T_a = 7.05\phi K I_a = 7.05 \times 0.1216 \times 26.6 = 22.8 \text{ ft-lb}$$

2-39 The mechanical efficiency of a shunt motor whose armature and field resistances are 0.055 and 32 ohms, respectively, is to be tested by means of a rope brake. When turning at 1400 rpm, the longitudinal pull on the 6-inch-diameter pulley is 57 lb. Simultaneous readings on the line voltmeter and ammeter are 105 volts and 35 amperes, respectively. Calculate (a) the counter-emf developed;(b) the copper losses;(c) the efficiency.

$$I_a = I_L - I_f = 35 - \frac{105}{32} - 31.7 \text{ amperes}$$

(a) $E_a = E_t - (I_a R_a) = 105 - (31.7 \times 0.055)$
$= 103.26$ volts

(b) $P_c = I_f^2 R_f + I_a^2 R_a$
$= (3.3^2 \times 32) + (31.7^2 \times 0.055)$
$= 404$ watts

(c) Output $= \dfrac{3p \times 1400 \times \frac{3}{12} \times 57}{33,000} = 3.8$ hp

Input $= \dfrac{105 \times 35}{746} = 4.93$ hp

$\eta_m = \dfrac{3.8}{4.93} = 0.771$, or 77%

2-40 A copper transmission line that is 1.5 miles in length is used to transmit 10 kilowatts from a 600-volt generating station. The voltage drop in the line is not to exceed 10% of the generating station voltage. Calculate (a) the line current;(b) the resistance of the line;(c) the cross-sectional area of the wire.

(a) $IL = \dfrac{10,000}{600} = 16.67$ amperes

The permissible voltage drop $= 600 \times 0.1 = 60$ volts

(b) $R = \dfrac{60}{16.67} = 3.6$ ohms

(c) $3.6 = \dfrac{10.4 \times 3 \times 5280}{A}$

$A = \dfrac{10.4 \times 3 \times 5280}{3.6} = 45,760$ CM

2-41 A trolley system 10 miles long is fed by two substations that generate 600 volts and 560 volts, respectively. The resistance of the trolley wire and rail return is 0.3 ohm per mile. If a car located 4 miles from the 600-volt substation draws 200 amperes from the line, what is the voltage between the trolley collector and the track? How much current is supplied by each substation?

With reference to Figure 2-11, the following equation can be written:

Equation (1) $I_1 + I_2 = 200$ amperes

That is, the arithmetical sum of the current drain from each sub-station must equal the current drawn by the trolley car. Similarly, the equations for the voltage drop in each branch of the trolley wire are:

Equation (2) $I_1(1.2) = 600 - E$
Equation (3) $I_2(1.8) = 560 - E$

Subtracting equation (3) from equation (2),

Equation (4) $40 = 1.2I_1 - 1.8I_2$

According to equation (1),

$$I_1 = 200 - I_2$$

Therefore, equation (4) becomes:

$$40 = 1.2(200 - I_2) - 1.8I_2$$
$$I_2 = 66.67 \text{ amperes}$$
$$I_1 = 200 - 66.67 = 133.33 \text{ amperes}$$

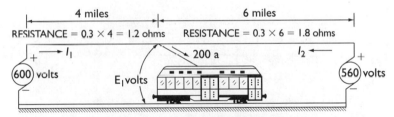

Figure 2-11   Current and potential drop in a trolley feeder system.

By inserting the value of I_2 in equation (3), obtain the voltage between the trolley collector and the track:

$$E = 560 - (1.8 \times 66.67) = 440 \text{ volts}$$

The same result can be obtained by inserting the value of I_1 in equation (2).

2-42 It is desired to supply power from a 220-volt source to points C and D in Figure 2-12 by means of the feeder arrangement indicated. The motor at point C requires 120 amperes, and the motor at point D requires 80 amperes. With the length of the wires as indicated and a maximum voltage drop of 10%, calculate (a) the cross-sectional area of feeder AB;(b) the cross-sectional area of feeder BC;(c) the cross-sectional area of feeder BD;(d) the power loss in each section.

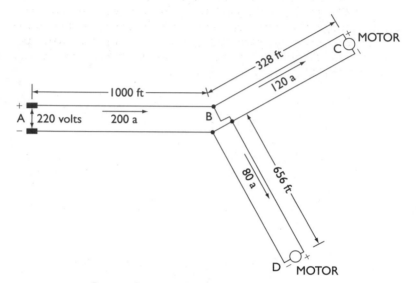

Figure 2-12 Branch feeder calculations.

The voltage drop across AC and AD is:

$$E' = 220 \times 0.1 = 22 \text{ volts}$$

To simplify our calculation, the voltage drop across BC and BD can be arbitrarily set at 10 volts. The voltage drop across AB is, therefore, 22.10, or 12 volts.

(a) $A = \left(\dfrac{10.4 \times 2}{12} \right) \times 200 \times 1000 = 346{,}667 \text{ CM}$

(b) $A = \left(\dfrac{10.4 \times 2}{10} \right) \times 120 \times 328 = 81{,}869 \text{ CM}$

(c) $A = \left(\dfrac{10.4 \times 2}{10} \right) \times 80 \times 656 = 109{,}158 \text{ CM}$

(d) $P_{AB} = 200 \times 12 = 2400$ watts

$P_{BC} = 120 \times 10 = 1200$ watts

$P_{BD} = 80 \times 10 = 800$ watts

2-43 The motor illustrated in Figure 2-13 is located at a distance of 500 feet from the generator and requires 40 amperes at 220 volts. No. 4 AWG wire is used. Calculate (a) the voltage at the generator; (b) the voltage-drop percentage in the line; (c) the power loss in the line; (d) the power-loss percentage; (e) the cost of power losses per year. Assume that the motor operates 8 hours per day, 300 days per year, at a cost of 3 cents per kilowatt-hour.

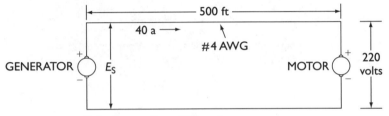

Figure 2-13 Calculations for a 220-volt motor.

With reference to Table 2-1, the cross-sectional area of No. 4 wire = 41,740 CM.

$$R = \frac{10.4 \times 1000}{41,740} = 0.25 \text{ ohms}$$

(a) $E_G = 220 + (40 \times 0.25) = 230$ volts

(b) $\dfrac{(230 - 220)100}{220} = 4.55\%$

(c) $P_G - P_R = 40^2 \times 0.25 = 400$ watts

(d) $\dfrac{(P_G - P_R)100}{P_R} = \dfrac{400 \times 100}{40 \times 220} = 4.55\%$

(e) Yearly cost of power losses = $0.4 \times 8 \times 300 \times 0.03$

$= \$28.80$

2-44 Energy is transmitted from a switchboard to the combined load shown in Figure 2-14. The lamp group requires 20 amperes, and the motor requires 30 amperes from the line. Number 2 wire (resistance = 0.156 ohm per 1000 ft) is used throughout the circuit. Calculate (a) the power drawn by the lamps; (b) the power drawn by the motor; (c) the power loss in the line; (d) the total power supplied by the switchboard.

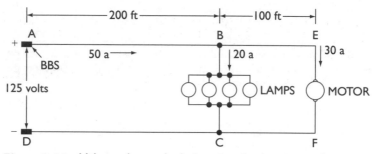

Figure 2-14   Voltage-drop calculations in a feeder circuit that is supplying a motor and lamp load.

The resistance in line ABCD is:

$$\frac{0.162 \times 200 \times 2}{1000} = 0.065 \text{ ohm}$$

(a) $P = 20 \times 121.8 = 2.436$ kilowatts

The resistance in line BEFC is:

$$\frac{0.162 \times 100 \times 2}{1000} = 0.0324 \text{ ohm}$$
$$\text{Voltage at the motor} = 121.8 - (30 \times 0.0324)$$
$$= 120.83 \text{ volts}$$

(a) $P_M = 30 \times 120.83 = 3.62$ kw

(b) $P_L = (50 \times 125) - (2436 + 3630) = 184$ watts

(c) $P_t = 125 \times 50 = 6.25$ kilowatts

Relative Conductivity

2-45 What is the comparison of electrical conductivity between silver, copper, and aluminum?

Of the three, silver is the best conductor of electricity and is considered to possess 100% conductivity. Copper is next, at approximately 94% of the conductivity of silver. Aluminum is considered to be last with approximately 61% of the conductivity of silver.

2-46 What are the advantages and disadvantages of silver as a conductor?

The price of silver makes it generally prohibitive to use for a conductor. It is used only in special cases for its high conductivity and where the price is a minor factor.

2-47 What are the advantages of copper as a conductor? What are its disadvantages?

Copper is plentiful, relatively inexpensive, a good conductor, and it has a high tensile strength. It can be corroded under certain circumstances.

2-48 What are the advantages and disadvantages of aluminum as a conductor?

Aluminum is inexpensive, lightweight, and readily available. It corrodes more easily than copper, so its use is limited to some degree; it also has less tensile strength than copper.

Chapter 3

Power and Power Factor

The unit of *work* is the foot-pound (ft-lb). This is the amount of work done when a force of one pound acts through a distance of one foot. The amount of work done is equal to the force in pounds times the distance in feet:

$$W \,(\text{work}) = \text{force} \times \text{distance}$$

Thus, if an object weighing 10 pounds is lifted a distance of 4 feet, the work done is equal to 40 ft-lb.

Time becomes involved when performing work, so you use the quantity known as *power*, which is the rate of doing work. Power is directly proportional to the amount of work done and is inversely proportional to the time in which the work is done. For example, more power is required if an object weighing 10 pounds is lifted through a distance of 4 feet in 1 minute than if the same 10-pound object is lifted a distance of 4 feet in 5 minutes. From this statement, you may arrive at the formula for power:

$$\text{Power (foot-pounds per minute)} = \frac{\text{work done (foot-pounds)}}{\text{time (minutes)}}$$

A more common unit of power is the *horsepower* (hp), which is equivalent to 33,000 ft-lb of work per minute. Remember this figure and its relationship to time and horsepower; it will be used quite often when working problems that deal with power.

In electricity, the unit of power is the watt. However, since the watt is a relatively small unit, the kilowatt is more commonly used as the unit of power. One kilowatt is equivalent to 1000 watts.

A good working knowledge of the electrical formulas that are used to determine power is a *must* for the electrician. When using electrical formulas to determine power, it is a universal practice to use the following notations:

P is the power in watts

I is the current in amperes

R is the resistance in ohms

E is the potential difference in volts

Thus, the power P expended in a load resistance R when a current I flows due to a voltage pressure E can be found by the following relationships:

$$P = IE$$
$$P = \frac{E^2}{R}$$
$$P = I^2R$$

Remember, IR equals a potential, in volts, and I^2R equals power, in watts.

When dealing with large amounts of electrical power, you may be required to determine the cost of the power consumed. You will be dealing with kilowatts and also with kilowatt-hours (kwh), which are the number of kilowatts used per hour. Thus, 25 kwh is 25 kilowatts used for 1 hour. To find the cost of electricity usage on a bill, use the following formula:

$$\text{Cost} = \frac{\text{watts} \times \text{hours used} \times \text{rate per kwh}}{1000}$$

For example, an electric heater that draws 1350 watts is used for 4 hours, and the cost of electricity for that particular location is 3 cents per kilowatt-hour. What is the cost of using the heater?

$$\text{Cost} = \frac{1350 \times 4 \times 0.03}{1000} = 16.2 \text{ cents}$$

Energy can be changed from one form to another but can never be destroyed. Therefore, you may readily change electrical power into mechanical power, and the converse is also true. The usual method of referring to mechanical power is in terms of horsepower: one horsepower is equal to 746 watts. This wattage value for the horsepower unit assumes that the equipment used to produce one horsepower operates at 100% efficiency. That, of course, is not possible, because there is always some power lost in the form of friction or other losses, which will be covered later in this text.

3-1 **A motor draws 50 amperes and is fed by a line of No. 6 copper wire that is 125 feet long. What is the I^2R loss of the line?**

In *Chapter 9, Table 8* of the NEC, No. 6 copper has a resistance of 0.410 ohm per 1000 feet at 25°C. The line is 125 feet long, so the amount of wire used will be 2×125, or 250 feet.

This is 25% of 1000 feet, so the resistance of the wire will be 0.410×0.25, or 0.1025 ohm. Therefore,

$$P = I^2R = (50)^2 \times 0.1025$$
$$= 2500 \times 0.1025 = 256.25 \text{ watts}$$

3-2 The line loss of the line in question 3-1 is 256.25 watts. If the motor is operated for 100 hours and the rate of electricity is 3 cents per kwh, what would be the cost of the I^2R loss of the line?

$$\text{Cost} = \frac{256.25 \times 100 \times 0.03}{1000} = 76.875 \text{ cents}$$

3-3 A 1-hp motor draws 1000 watts. What is its efficiency?

$$\text{Efficiency} = \frac{\text{output}}{\text{input}}$$

Therefore,

$$\text{Efficiency} = \frac{746}{1000} = 0.746, \text{ or } 74.6\%$$

3-4 An electric iron draws 11 amperes at 120 volts. How much power is used by the iron?

$$P = 11 \times 120 = 1320 \text{ watts} = 1.32 \text{ kw}$$

3-5 A motor must lift an elevator car weighing 2000 pounds to a height of 1000 feet in 4 minutes. (a) What is the theoretical size, in horsepower, of the motor required? (b) At 50% efficiency, what is the size, in horsepower, of the motor required?

(a) $W = 2000 \times 1000 = 2,000,000 \text{ ft-lb}$

$$\frac{2,000,000}{4} = 500,000 \text{ ft-lb per minute}$$

$$\frac{500,000}{33,000} = 15.15 \text{ hp}$$

(b) $\text{Input} = \dfrac{\text{output}}{\text{efficiency}} = \dfrac{15.15}{0.50} = 30.3 \text{ hp}$

A 30-hp motor will carry this load nicely.

3-6 A lamp operating at 120 volts has a resistance of 240 ohms. What is the wattage of the lamp?

$$P = \frac{E^2}{R} = \frac{120^2}{240} = \frac{14,400}{240} = 60 \text{ watts}$$

3-7 What is the overall efficiency of a 5-hp motor that draws 20 amperes at 240 volts?

$$\text{Input} = 240 \times 20 = 4800 \text{ watts}$$
$$\text{Output} = 5 \times 746 = 3730 \text{ watts}$$
$$\text{Efficiency} = \frac{3750}{4800} = 0.777, \text{ or } 77.7\%$$

3-8 What is the cost of operating a 2-watt electric clock for one year at 2 cents per kwh?

$$\text{Cost} = \frac{2 \text{ watts} \times 24 \text{ hours} \times 365 \text{ days} \times 0.02}{1000}$$
$$= 35.04 \text{ cents}$$

3-9 The primary of a transformer draws 4 amperes at 7200 volts. A reading at the secondary shows 110 amperes at 240 volts. What is the efficiency of the transformer at this load?

$$\text{Efficiency} = \frac{\text{output}}{\text{input}} = \frac{110 \times 240}{4 \times 7200}$$
$$= \frac{26,400}{28,800} = 91.67\%$$

3-10 What instrument is used to measure voltage?
A voltmeter.

3-11 How is a voltmeter connected in a circuit? (explain and illustrate)
A voltmeter is always connected in shunt, or parallel, across the load or the source whose voltage is being measured. This is illustrated in Figure 3-1.

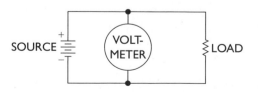

Figure 3-1 Voltmeter connections in a circuit.

3-12 With what instrument do you measure current?
An ammeter.

3-13 How is an ammeter connected in a circuit? (explain and illustrate)
An ammeter is connected in series with the circuit being tested. This is illustrated in Figure 3-2. Clamp-on ammeters, however, operate by measuring the magnetic field around conductors. These meters, though less accurate, are not wired into the circuit at all. Rather, they are simply clamped around the conductor whose current is to be measured.

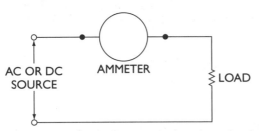

Figure 3-2 Ammeter connections in a circuit.

3-14 Give two methods of measuring large currents. Explain why they must be used.
It is impractical to construct a meter that is capable of carrying the large currents. Therefore, when measuring large currents in ac or dc circuits the electrician uses a shunt, which consists of a low-resistance load connected in series with the load and connected in parallel with a high-resistance meter. The meter then receives only a small fraction of the current passing through the load. This method is illustrated in Figure 3-3.

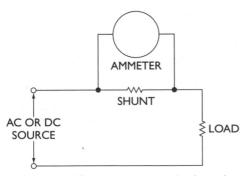

Figure 3-3 One common method used to measure large current.

Another method for measuring large currents in an ac circuit makes use of a current transformer (Figure 3-4), which is described in Chapter 6. You may also use the wire that carries the load current as the primary of a current transformer, in conjunction with a clip-on ammeter. The secondary of the transformer is incorporated into the ammeter, so the meter has no actual physical connection in the circuit (Figure 3-4).

3-15 When using an ammeter, what precautions must be taken?
Know whether the current is ac or dc, and use the appropriate ammeter. Be sure that the rating of the meter is large enough for the current being measured to prevent the meter from being damaged.

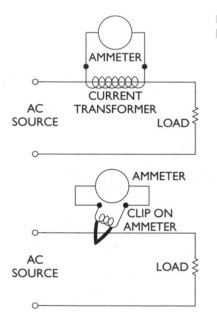

Figure 3-4 Measurement of large currents in an AC circuit.

3-16 What is a wattmeter?
A wattmeter is an instrument that is designed to measure directly the active power in an electric circuit. It consists of a coil connected in series with the circuit, such as in an ammeter, and a coil connected in parallel with the circuit, such as in a voltmeter. Both coils actuate the same meter, thereby measuring both the current and the voltage affecting one meter, which may be calibrated in watts, kilowatts, or megawatts.

3-17 How is a wattmeter connected in a circuit? Illustrate.
See Figure 3-5.

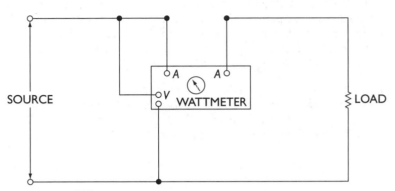

Figure 3-5   Wattmeter connections in a circuit.

3-18 Can the principles used in dc circuits be applied to all ac circuits?
The fundamental principles of dc circuits may also apply to ac circuits that are strictly resistive in nature, such as incandescent lighting and heating loads.

3-19 What causes inductive reactance?
Inductive reactance is caused by opposition to the flow of alternating current by the inductance of the circuit.

3-20 Give some examples of equipment that causes inductive reactance.
Motors, transformers, choke coils, relay coils, ballasts.

3-21 When only inductive reactance is present in an ac circuit, what happens to the current in relation to the voltage?
The current is said to *lag* behind the voltage.

3-22 What is the reason for current lagging the voltage in an inductive circuit?
In an ac circuit, the current is continually changing its direction of flow (60 times a second in a 60-hertz circuit). Any change of current value is opposed by the inductance within the circuit involved.

3-23 Draw sine waves of voltage and current in a circuit containing only inductive reactance when 60-hertz ac is applied to the circuit.
See Figure 3-6.

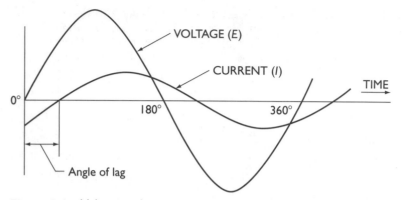

Figure 3-6 Voltage and current sine waves in a circuit containing only inductive reactance, with a frequency of 60 hertz.

3-24 When only capacitive reactive is present in an ac circuit, what happens to the relationship that exists between the voltage and the current?

The current is said to *lead* the voltage.

3-25 Draw sine waves of current and voltage in a circuit containing only capacitive reactance when 60-hertz ac is applied to the circuit.

See Figure 3-7.

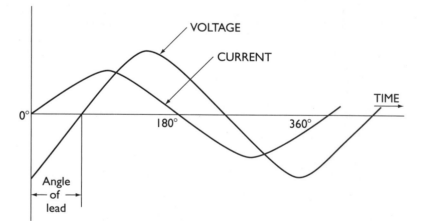

Figure 3-7 Voltage and current sine waves in a circuit containing only capacitive reactance, with a frequency of 60 hertz.

3-26 In a purely resistive circuit with an applied 60-hertz ac voltage, what is the relationship between the current and voltage?
The current and voltage will be in phase, as illustrated in Figure 3-8.

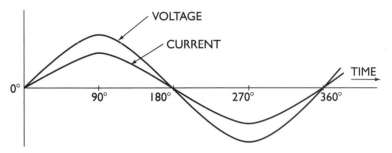

Figure 3-8 Voltage-current relationship in a resistive circuit at a frequency of 60 hertz.

3-27 Is it possible to have a circuit with only inductive reactance?
Under ordinary circumstances this is not possible, since all metal conductors have a certain amount of resistance. It is, however, possible for circuits that employ superconductors. At present, this can be done only in laboratory situations.

3-28 If it were possible to have only inductive reactance in an ac circuit, by what angle would the current lag the voltage?
The current would lag the voltage by 90°. When the voltage was at its maximum value in one direction, the current would be zero and would be just getting ready to increase in the direction of maximum voltage.

3-29 If it were possible to have only capacitive reactance in an ac circuit, by what angle would the current lead the voltage?
The current would lead the voltage by 90°. When the current was at its maximum value in one direction, the voltage would be zero and would be just getting ready to increase in the direction of maximum current.

3-30 Draw vectorially the current and voltage 90° out of phase in an inductive circuit.
See Figure 3-9.

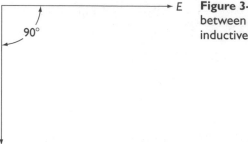

Figure 3-9 Vector relationship between voltage and current in an inductive circuit.

3-31 Draw vectorially the current and voltage 90° out of phase in a capacitive circuit.

See Figure 3-10.

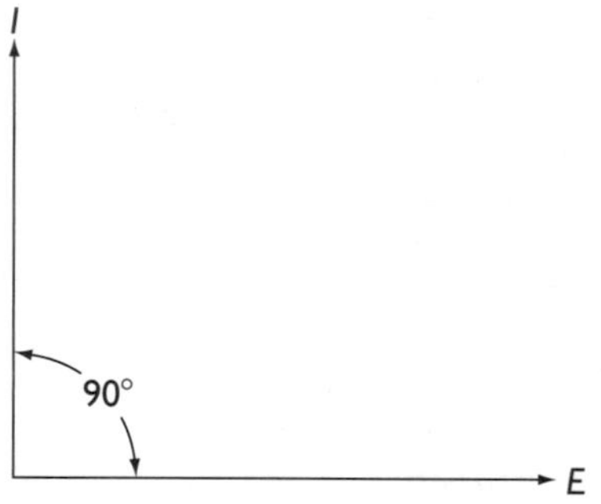

Figure 3-10 Vector relationship between voltage and current in a capacitive circuit.

3-32 What is meant by power factor?

Power factor is the phase displacement of current and voltage in an ac circuit. The cosine of the phase angle of displacement is the power factor; the cosine is multiplied by 100 and is expressed as a percentage. The cosine of 90° is 0; therefore, the power factor is 0%. If the angle of displacement were 60°, the cosine of which is 0.500, the power factor would be 50%. This is true whether the current leads or lags the voltage.

3-33 How is power expressed in dc circuits and ac circuits that are purely resistive in nature?

In dc circuits and ac circuits that contain only resistance,

$$P \text{ (watts)} = E \times I$$
$$P \text{ (kilowatts)} = \frac{E \times I}{1000}$$

3-34 How is power expressed in ac circuits that contain inductive and/or capacitive reactance?

$$VA \text{ (volt-amperes)} = E \times I$$
$$KVA \text{ (kilovolt-amperes)} = \frac{E \times I}{1000}$$
$$P \text{ (watts)} = E \times I \times \text{power factor}$$

3-35 In a 60-Hz ac circuit, if the voltage is 120 volts, the current is 12 amperes, and the current lags the voltage by 60°, find (a) the power factor; (b) the power in volt-amperes (VA); (c) the power in watts.

(a) The cosine of 60° is 0.500; therefore, the power factor is 50%.

(b) 120 × 12 = 1440 VA, which is called the *apparent* power.

(c) 120 × 12 × 0.5 = 720 watts, which is called the *true* power.

3-36 Show vectorially a 12-ampere line current lagging the voltage by 60°; indicate that the in-phase current is 50%, or 6 amperes.
See Figure 3-11.

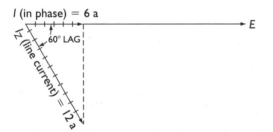

Figure 3-11 A 12-ampere line current lagging the voltage by 60°; the in-phase current is 50% of the line current, or 6 amperes.

3-37 Why is a large power factor of great importance?
As can be seen in questions 3-35 and 3-36, there is an apparent power of 1440 VA and a true power of 720 watts. There are also 12 amperes of line current and 6 amperes of in-phase, or effective, current. This means that all equipment from the source of supply to the power-consumption device must be capable of handling a current of 12 amperes, while actually the device is only utilizing a current of 6 amperes. A 50% power factor was used intentionally to make the results more pronounced. The I^2R loss is based on the 12-ampere current, whereas only 6 amperes are really effective.

3-38 How can power factor be measured or determined?
There are two easy methods: (1) by the combined use of a volt-meter, ammeter, and wattmeter (all ac instruments, of course), and (2) by the use of a power-factor meter.

3-39 How is the voltmeter-ammeter-wattmeter method used to determine the power factor?
The voltmeter, ammeter, and wattmeter are connected properly in the circuit. Then the readings of the three meters are simultaneously taken under the same load conditions. Finally, the following calculations are made:

$$\text{Power factor} = \frac{\text{true power (watts)}}{\text{apparent power } (E \times I)}$$
$$= \frac{\text{w}}{E \times I}$$
$$= \frac{\text{kw}}{\text{KVA}}$$

3-40 What is a power-factor meter?
A power-factor meter is a wattmeter calibrated to read the power factor directly, instead of in watts. It is connected in the circuit in the same manner as a wattmeter.

3-41 In a circuit that contains an inductance, such as a motor, will the current lead or lag the voltage? What steps can be taken to correct the power factor?
The current will lag the voltage. The power factor can be corrected by adding capacitance to the circuit. This can be done either by introducing capacitors to the circuit, or by using synchronous motors and overexciting their fields.

3-42 What is the ideal power transmission condition, as far as the power factor is concerned?
Unity power factor, or a power factor of 1, which means that the current is in phase with the voltage.

3-43 Which is the best condition, a leading or a lagging current?
They both have the same effect, so one is as good as the other. In an overall picture, there are usually more lagging currents

than leading currents, because most of the loads used in the electrical field are of an inductive nature.

3-44 In an ac voltage or current, there are three values that are referred to. What are they?
The *maximum* value of current or voltage, the *effective* value of current or voltage, and the *average* value of current or voltage.

3-45 What is meant by the *maximum* ac voltage?
This is the maximum, or peak, voltage value of an ac sine wave.

3-46 What is meant by the *average* ac voltage?
This is the average of the voltages taken at all points on the sine wave.

3-47 What is meant by the *effective* ac voltage?
This is the value of the useful voltage that is indicated on a voltmeter. It is the voltage that is used in all normal calculations in electrical circuits.

3-48 What percentage of the peak voltage is the effective voltage?
70.7% of the peak value.

3-49 What percentage of the peak voltage is the average voltage?
63.7% of the peak voltage.

3-50 What is the effective voltage most commonly called?
Effective voltage is usually called rms voltage. (The term rms means rootmean-square.)

3-51 How could the rms voltage be arrived at vectorially?
An infinite number of lines could be drawn from the base line to the one-half amplitude value of the sine wave. These voltage values would then be squared, the sum of the squares would be averaged (added together and divided by the number of squares), and the square root of this figure would be the rms value of the voltage in question.

3-52 Does the peak voltage have to be considered?
Yes. In design, the peak voltage must be considered, because this value is reached twice in every cycle.

3-53 Draw a sine wave showing the effective value and the peak value of a standard 120-volt ac line.

See Figure 3-12.

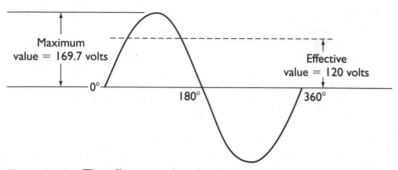

Figure 3-12   The effective and peak values of a standard 120-volt line.

Chapter 4

Lighting

4-1   What is a grounded (neutral) conductor?

A white or natural gray conductor or a conductor with three longitudinal white stripes, which indicates that it is the grounded circuit conductor.

4-2   May a grounded conductor ever be used as a current-carrying conductor? Explain.

It is permissible to use a white or natural gray conductor as a current-carrying conductor, if the conductor is permanently re-identified by painting or other effective means at each location where the conductors are visible and accessible [see *Section 200.7(A)* of the NEC]. Note, however, that when used as just described, the white or gray conductor is removed from ground and is no longer a grounded conductor.

4-3   Draw a diagram of the proper connections for a two-wire cable, one white wire and one black, to supply a light from a single-pole switch.

See Figure 4-1.

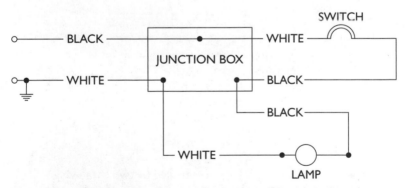

Figure 4-1   A two-wire cable supplying a lamp from a single-pole switch.

4-4   When must the neutral, or grounded, conductor be provided with a switch?

A lamp or pump circuit on a gasoline dispensing island *must* provide both the neutral and the hot wires, which supply the light or the pump, with a switch (see NEC, *Section 514.11(A)*).

4-5 Draw a three-way switch system and a lamp, using cable; show the colors and connections.
See Figure 4-2.

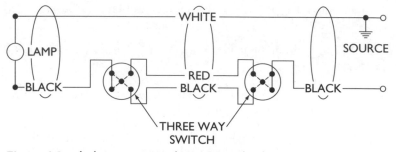

Figure 4-2 A three-way switch system with a lamp.

4-6 Draw a circuit supplying a lamp that is controlled by two three-way switches and one four-way switch (use cable).
See Figure 4-3.

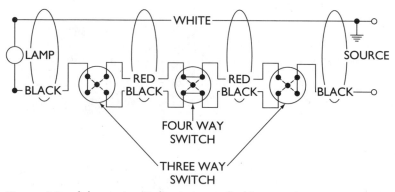

Figure 4-3 A lamp circuit that is controlled by two three-way switches and one four-way switch.

4-7 Draw a master-control lighting system.
See Figure 4-4.

4-8 Draw an electrolier switching circuit for controlling lights.
See Figure 4-5.

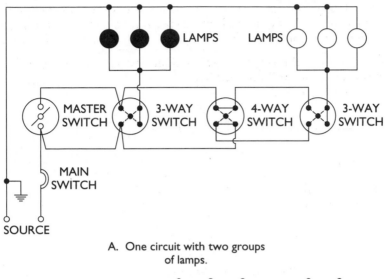

A. One circuit with two groups
of lamps.

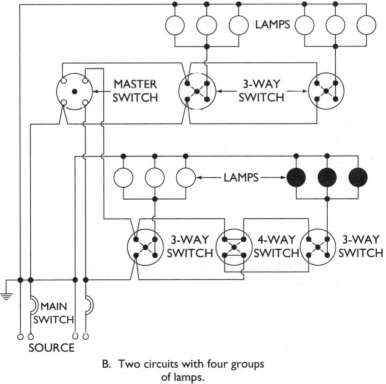

B. Two circuits with four groups
of lamps.

Figure 4-4   A master-control lighting system.

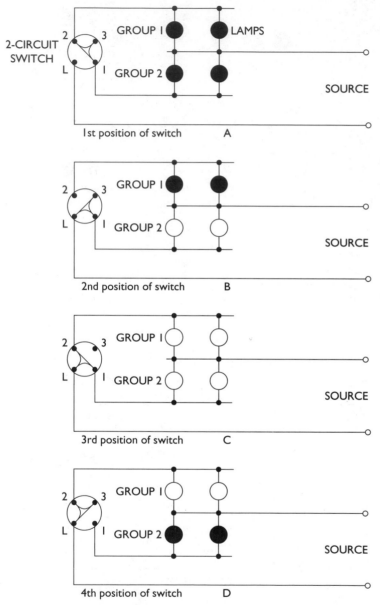

Figure 4-5 An electrolier switch arrangement for control of lamp circuits.

4-9 Draw an electrolier switching circuit to control three sets of lamps.

See Figure 4-6.

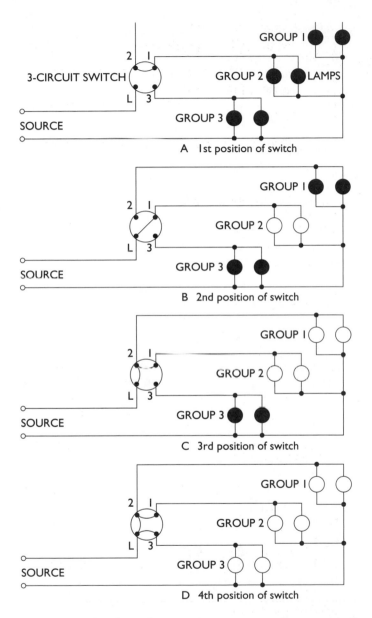

Figure 4.6 An electrolier switch arrangement for control of three groups of lamps. The sequence of operation is depicted diagrammatically.

4-10 What is a preheated-type fluorescent tube?

There are two external contacts at each end of a glass tube. Each set of contacts is connected to a specially treated tungsten filament. The inside of the tube is coated with a fluorescent powder, the type of powder used controls the color output of the tube. The tube is filled with an inert gas, such as argon, and a small drop of mercury to facilitate starting.

4-11 What kind of light does the fluorescent bulb produce within the bulb itself?

Ultraviolet light.

4-12 Do the filaments stay lit during the operation of a fluorescent bulb? Explain.

No. They remain lit only at the start to vaporize the mercury; they are then shut off by the starter. Current is supplied to one contact on each end, thereby sustaining the mercury arc within the tube.

4-13 Draw a simple circuit of a fluorescent lamp and fixture.

See Figure 4-7.

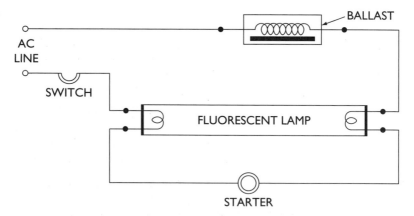

Figure 4-7 A fluorescent lamp and fixture circuit.

4-14 Explain the action of the glow-type starter.

When the switch is closed, the high resistance in the glow bulb of the starter produces heat, which causes the bimetallic U-shaped strip to close the contacts, thereby lighting the filaments in the fluorescent bulb. When the contacts close, the glow bulb cools and allows the contacts to open, thus disconnecting one side of the filaments. The arc within the fluorescent bulb is sustained without keeping the filaments heated, and the bulb lights.

4-15 Is the power factor of the circuit in Figure 4-7 good or bad? Is the current leading or lagging the voltage?

The power factor is bad; the current is lagging the voltage because of the inductive reactance of the ballast.

4-16 How can the power factor of the circuit in Figure 4-7 be corrected?

By the addition of capacitors in the ballast to counteract the lagging power factor.

4-17 What is the fixture in question 4-16 called?

A power-factor-corrected fixture.

4-18 What is a trigger-start fluorescent fixture?

This fixture uses a special trigger-start ballast that automatically preheats the filaments without the use of a starter.

4-19 Is a fluorescent fixture more efficient than an incandescent fixture?

Yes. However, no fixed efficiency can be quoted because it varies with the size of the bulb, the ballast, etc. As a rule of thumb, a 40-watt fluorescent bulb is generally considered to put out about the same amount of light as a 100-watt incandescent bulb.

4-20 What is an instant-start Slimline fluorescent lamp?

This lamp has a single terminal at each end. The ballast is normally of the autotransformer type, which delivers a high voltage at the start and a normal voltage after the arc is established.

4-21 Does room temperature affect the operation of fluorescent lamps?

Yes. The normal fluorescent lamp is designed for operation at 50°F or higher.

4-22 If the operating temperature is expected to be below 50°F, what measures should be taken when using fluorescent lamps?

Use special lamps and starters that are designed for lower temperatures.

4-23 Do fluorescent lamps produce a stroboscopic effect? Explain.
Yes. This effect is due to the fact that at 60 hertz the current passes through zero 120 times a second.

4-24 How can the strobe effect be compensated for in a two-bulb system?
When the power factor is corrected, the capacitor is so connected that at the instant the current in one bulb is passing through zero, the current in the other bulb is not at zero, and the strobe effect goes unnoticed.

4-25 What is another way to minimize the strobe effect?
By using a higher frequency, such as 400 hertz.

4-26 Does a 400-hertz frequency have any other advantages?
Yes. Smaller and lighter ballasts can be used. This makes it feasible to use simple capacitance-type ballasts, which produce an overall gain in efficiency.

4-27 Does frequent starting and stopping of fluorescent lighting affect the bulb life?
The life of fluorescent bulbs is affected by starting and stopping. A bulb that is constantly left on will have a much longer life than one that is turned on and off frequently.

4-28 When lighting an outdoor activity area with large incandescent bulbs, how can the light output of the bulbs be increased?
By using bulbs of a lower voltage rating than that of the source of supply. For example, by using a bulb that is rated at 105 volts on a 120-volt supply, the output can be increased by roughly 30%, but the life of the bulb will be cut by about 10%.

4-29 How is the light output of lamps rated?
In lumens.

4-30 How are light levels rated?
In foot-candles.

4-31 What is a foot-candle?
The amount of direct light emitted by one international candle on a square foot of surface, every part of which is one foot away.

Chapter 5

Branch Circuits and Feeders

5-1 Is the NEC a law?
It is not a law, but it is adopted into laws that are established by governmental agencies; it may be adopted in its entirety, in part, or with amendments.

5-2 Who has the responsibility for Code interpretations?
The administrative authority that has jurisdiction over endorsement of the Code has the responsibility for making Code interpretations (see NEC, *Section 90.4* and *Section 80.2*). Usually, this is a local electrical inspector.

5-3 When are rules in the NEC mandatory and when are they advisory?
When the word "shall" is used, the rules are mandatory. Fine Print Notes (FPN) explain the intent of Code rules (see *Section 9.5* of the Code).

5-4 According to the NEC, what are voltages?
Throughout the Code, the voltage considered shall be that at which the circuit operates. This is often expressed as, for example, 600 volts, nominal (see NEC, *Section 220.2*).

5-5 When wire gage or size is referred to, what wire gage is used?
The American Wire Gage (AWG) or in circular mils (CM).

5-6 When referring to conductors, what material is referred to?
Copper, unless otherwise specified. When other materials are to be used, the wire sizes must be changed accordingly (see NEC, *Section 110.5*).

5-7 In what manner must the work be executed?
All electrical equipment must be installed in a neat and workmanlike manner (see NEC, *Section 110.12*).

5-8 May wooden plugs be used for mounting equipment in masonry, concrete, plaster, etc.?
No (see NEC, *Section 110.13*).

5-9   How must conductors be spliced or joined together?

They must be spliced or joined together by approved splicing devices or by brazing, welding, or soldering with a fusible metal or alloy (see NEC, *Section 110.14*).

5-10   When soldering, what precautions must be used?

All joints or splices must be electrically and mechanically secure before soldering and then soldered with a noncorrosive flux (see NEC, *Section 110.14*). This does not apply to conductors for grounding purposes; soldering is not allowed on these conductors.

5-11   How should splices or joints be insulated?

They must be covered with an insulation that is equivalent to the original conductor insulation (see NEC, *Section 110.14*).

5-12   Can an autotransformer be used on an ungrounded system?

The autotransformer must have a grounded conductor that is common to both primary and secondary circuits and tied into a grounded conductor on the system supplying the autotransformer (see NEC, *Section 210.9*).

An autotransformer may be used to extend or add an individual branch circuit in an existing installation for equipment load without the connection to a similar grounded conductor when transforming from a nominal 208-volt supply or similarly from 240 volts to 208 volts.

5-13   On No. 6 or smaller conductors, what means must be used for the identification of the grounded conductors?

Insulated conductors of No. 6 or smaller, when used as grounded conductors, must be white, natural gray, or colored (but never green) with three longitudinal white stripes. On Type MI cable, the conductors must be distinctively marked at the terminal during installation (see NEC, *Section 200.6*).

5-14   How should conductors larger than No. 6 be marked to indicate the grounded wire?

By the use of white, natural gray, or colored (but never green) insulation with three longitudinal white stripes, or by identifying with a distinctive white marking at the terminals during installation (see NEC, *Section 200.6*).

5-15   How is the high-leg conductor of a 4-wire delta identified?

When the midpoint of one phase is grounded to supply lighting and similar loads on a 4-wire delta-connected secondary, the phase conductor with the higher voltage to ground shall be orange in color or be indicated by tagging or other effective means at the point where a connection is to be made if the neutral conductor is present (see *Section 110.15* of the NEC).

5-16 On a grounded system, which wire must be connected to the screw shell of a lampholder?
The grounded conductor [see NEC, *Section 200.10(C)*].

5-17 What will determine the classification of branch circuits?
The maximum permitted setting or rating of the overcurrent-protective device in the circuit (see NEC, *Section 210.20*).

5-18 What color-coding is required on multiwire branch circuits?
The grounded conductor of a branch circuit shall be identified by a continuous white or natural gray color. Whenever conductors of different systems are installed in the same raceway, box, auxiliary gutter, or other types of enclosures, one system grounded conductor, if required, shall have an outer covering of white or natural gray. Each other system grounded conductor, if required, shall have an outer covering of white with an identifiable colored stripe (not green) running along the insulation or another means of identification. Ungrounded conductors of different voltages shall be a different color or identified by other means (see NEC, *Section 200.6(D)*).

5-19 How must a conductor that is used only for equipment and grounding purposes be identified?
By the use of a green color, or green with one or more yellow stripes, or by being bare (only grounding conductors may be bare).

5-20 Can green-colored wire be used for circuit wires?
No. Green is intended for identification of equipment-grounding conductors only (see NEC, *Sections 210.5(B)* and *250.119*).

5-21 What voltage is used between conductors that supply lampholders of the screw-shell type, receptacles, and appliances in dwellings?
Generally speaking, a voltage of 120 volts between conductors is considered the maximum. There are, however, some exceptions (see NEC, *Section 210.6*).

5-22 What are the exceptions to the 120-volt maximum between conductors?

Listed high-intensity discharge lighting fixtures with medium base lampholders, fixtures with mogul base lampholders, other types of lighting equipment (without screw-shell lampholders), utilization equipment that is either permanently connected or cord-and-plug connected (see NEC, *Section 210.6(C)* and *Section 80.2*).

5-23 How must you ground the grounding terminal of a grounding-type receptacle?

By the use of an equipment-grounding conductor of green covered wire, green with one or more yellow stripes, or bare conductors. However, the armor of Type AC metal-clad cable, the sheath of MI cable, or a metallic raceway is acceptable as a grounding means. *Section 250.118* does not permit the general use of flexible metal conduit as a grounding means unless it and the connectors are listed.

5-24 How can you ground the grounding terminal on a grounding-type receptacle on extensions to existing systems?

Run the grounding conductor to a grounded water pipe near the equipment (see NEC, *Sections 250.130* and *250.130(C)*).

5-25 What is the minimum size for branch-circuit conductors?

They cannot be smaller than No. 8 for ranges of 8¾ kW or higher rating and not smaller than No. 14 for other loads (see NEC, *Section 210.19(A)* and *(B)*).

5-26 What is the requirement concerning all receptacles on 15- and 20-ampere branch circuits?

All receptacles on 15- and 20-ampere branch circuits must be of the grounding type. A single receptacle installed on an individual branch circuit must have a rating of not less than the rating of the branch circuit. (For complete details, see NEC, *Section 210.7(A)* and *(B)*.)

5-27 What are the requirements for spacing receptacles in dwelling occupancies?

All receptacles in kitchens, family rooms, dining rooms, living rooms, parlors, libraries, dens, sun rooms, recreation rooms, and bedrooms must be installed so that no point along the wall space, measured horizontally, is more than 6 feet from a receptacle. This includes wall space that is 2 feet or wider and any space occupied by sliding panels on exterior walls. Sliding glass panels are excepted. At least one outlet must be installed for the laundry. (See

NEC, *Section 210.52.*) The wall space afforded by fitted room dividers, such as freestanding bar-type counters, must be included in the 6-foot measurement.

In the kitchen and dining areas, a receptacle outlet must be installed at each counter space that is 12 inches wide or more. Countertop spacers separated by range tops, refrigerators, or sinks must be considered as separate countertop space. Receptacles rendered inaccessible by the installation of appliances fastened in place or appliances occupying dedicated space won't be considered as the required outlets.

At least one wall receptacle outlet must be installed in the bathroom adjacent to each basin location.

5-28 In residential occupancies, will ground-fault circuit interrupters be required?

Yes, for all 125-volt, 15- and 20-ampere receptacle outlets installed out-of-doors for residential occupancies and also for receptacle outlets in bathrooms, basements, kitchens, and garages (see NEC, *Section 210.8*).

5-29 A bedroom has one wall that contains a closet 8 feet in length, with sliding doors and a wall space of 2 feet. The bedroom door opens into the room and back against this wall space. Where are receptacles required on this wall?

One receptacle is required in the 2-foot space (Figure 5-1). (See NEC, *Section 210.52(A)*.)

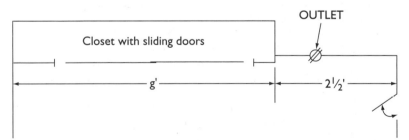

Figure 5-1 Receptacle requirements on a wall containing a closet with sliding doors.

5-30 Sketch a typical living room with sliding panels on the outside wall. Locate receptacles and show the room dimensions (Figure 5.2).

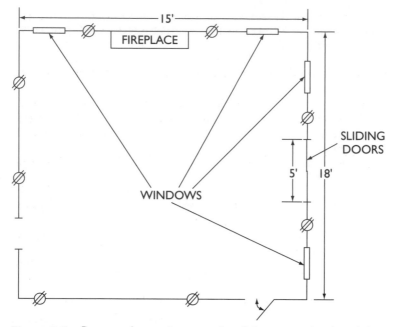

Figure 5-2 Receptacle requirements in a living room that has sliding doors on the outside wall.

5-31 A fastened-in-place appliance is located on a 15- or 20-ampere branch circuit on which there is a lighting fixture. What is the maximum rating that would be permitted on the appliance?

50 percent of the branch circuit rating (see NEC, *Section 210.23(A)*).

5-32 A cord-and-plug-connected appliance is used on a 15- or 20-ampere branch circuit. What is the maximum rating permitted for the appliance?

80 percent of the branch circuit rating (see NEC, *Section 210.23(A)*).

5-33 What is the smallest size wire permissible for a feeder circuit?
No. 10 wire (see NEC, *Section 215.2(B)*).

5-34 What are the permissible voltage drops allowable on feeders and branch circuits?

On feeders, not more than 3 percent for power, heating, and lighting loads or combinations thereof, and a total maximum voltage drop not to exceed 5 percent for conductors and for combinations of feeders and branch circuits (see NEC, *Section 215.2(D)*).

5-35 What is the basis for figuring the general lighting loads in occupancies?
They are figured on a volt-amperes-per-square-foot basis, using NEC. *Table 220.3(A)* can be used to determine the unit load per square foot in volt-amperes for the types of occupancies listed.

5-36 What measurements are used to determine the number of watts per square foot?
The outside dimensions of the building and the number of floors, not including open porches or garage (see NEC, *Section 200.3(A)*).

5-37 What is the unit load per square foot (in volt-amperes) for a hospital?
2 volt-amperes (see NEC, *Table 220.3(A)*).

5-38 What is the unit load per square foot (in volt-amperes) for a school?
3 volt-amperes (see NEC, *Table 220.3(A)*).

5-39 How many volt-amperes per square foot are required to be included for general-purpose receptacle outlets when a set of plans does not show their locations in an office building?
1 volt-ampere per square foot (see NEC, *Table 220.3(A)*** note).

5-40 What is the unit load per square foot (in volt-amperes) for a warehouse used for storage?
¼ volt-ampere (see NEC, *Table 220.3(A)*).

5-41 What voltages are used for purposes of computing feeder and branch-circuit loads when other voltages are not specified?
120, 120/240, 208Y/120, 240, 480Y/277, 480, and 600 volts (see NEC, *Section 220.2*).

5-42 How are continuous and noncontinuous loads for feeder ratings calculated?
Branch circuits must be rated no less than the noncontinuous load, plus at least 125 percent of the continuous load (see NEC, *Section 215.2*).

5-43 Are unfinished basements of dwellings used in figuring watts per square foot?
Yes, if adaptable for future use. If these spaces are not adaptable, they are not used (see NEC, *Section 220.3(A)*).

5-44 What loads are used in figuring outlets for other than general illumination?
Outlets supplying specific loads and appliances must use the ampere rating of the appliance. Outlets supplying heavy-duty lampholders must use 600 volt-amperes; calculations for other outlets must use 180 volt-amperes (see NEC, *Section 220.3(C)*).

5-45 What load must be figured for show-window lighting?
Not less than 200 volt-amperes for each linear foot measured horizontally along the base (see NEC, *Section 220.12*).

5-46 What are the receptacle requirements in dwelling occupancies for kitchen, family room, laundry, pantry, dining room, and breakfast room?
There shall be a minimum of two 20-ampere small-appliance circuits for the kitchen, family room, pantry, dining room, and breakfast room. There should also be a minimum of one 20-ampere circuit for the laundry (see NEC, *Section 220.16*).

5-47 What is the unit load per square foot (in volt-amperes) for a store?
3 volt-amperes (see NEC, *Table 220.3(A)*).

5-48 Outlets for heavy-duty lampholders are to be based on a load of how many volt-amperes?
600 (see NEC, *Section 220.3(B)(5)*).

5-49 How is the load for a household electric clothes dryer computed?
See NEC, *Section 220.18*.

5-50 Are demand factors permitted in determining feeder loads?
Yes, they are given in NEC, *Table 220.11*.

5-51 What size feeders must be installed in dwelling occupancies?
The computed load of a feeder must not be less than the sum of all branch circuit loads supplied by the feeder. The demand factors can be used in the calculation of feeder sizes.

5-52 When figuring the neutral load to electric ranges, what is the maximum unbalanced load considered to be?

The maximum unbalanced load for electric ranges is considered to be 70 percent of the load on the ungrounded conductors (see NEC, *Section 220.22*).

5-53 How is the neutral load on a 5-wire, 2-phase system determined?

It is figured at 140 percent of the load on the ungrounded conductors (see NEC, *Section 220.22*).

5-54 How may the neutral-feeder load on a 3-wire dc or single-phase 3-wire ac system be determined?

The 70 percent demand factor may be used on range loads, and a further demand factor of 70 percent may be used on that portion of the unbalanced load in excess of 200 amperes (see NEC, *Section 220.22*).

5-55 How do you calculate the unbalanced load on a 4-wire, 3-phase system?

The 70 percent demand factor may be used on range loads, and a further demand factor of 70 percent may be used on that portion of the unbalanced load in excess of 200 amperes (see NEC, *Section 220.22*).

5-56 Can you make a reduction on the neutral feeder load where discharge lighting, data processing, or similar equipment is involved?

No reduction can be made on the neutral capacity for the portion of the load that consists of electric discharge lighting, data processing, or similar equipment. The load on the neutral feeder must be taken at 100 percent of the ungrounded conductors (see NEC, *Section 220.22*).

5-57 Why do discharge lighting loads require no reduction in neutral feeder capacity?

Because of the effect of the third harmonic on the current value in the neutral feeder. In fact, it's sometimes necessary to oversize the neutral in such circuits. The same effect occurs when a great deal of data processing equipment is connected to the circuits.

5-58 Are demand factors applicable to electric ranges?

Yes. *Table 220.19* of the NEC can be used in determining demand factors for electric ranges.

5-59 Can demand factors be used on electric-clothes-dryer loads in the same manner as they are used on electric ranges?

Yes (see NEC, *Table 220.18*).

5-60 Are feeder demand factors permitted for commercial ranges and other commercial kitchen equipment?
    Yes (see NEC, *Table 220.20*).

5-61 There are 2500 square feet of floor area in a house that contains an electric range rated as 12 kW and an electric dryer rated at 5000 watts. Calculate (a) the general lighting required, (b) the minimum number and sizes of branch circuits required, and (c) the minimum size of feeders (service conductors) required.

(a) 2500 square feet at 3 volt-amperes per square foot equals 7500 watts.

(b) 7500 watts divided by 120 volts equals 62.5 amperes; this would require a minimum of five 15-ampere circuits with a minimum of No. 14 wire, or a minimum of four 20-ampere circuits with a minimum of No. 12 wire.

Small-appliance load: A minimum of two small-appliance circuits (see NEC, *Section 220.4(B)*) of 1500 volt-amperes each. These will require 20-ampere 2-wire circuits with No. 12 wire.

A 12-kW range has a demand of 8 kW, according to NEC, *Table 220-19*, and 8000 watts divided by 240 volts equals 33 amperes. Therefore, a minimum of No. 8 wire with a 40-ampere service would be required, although good practice would indicate the use of No. 6 wire with a 50-ampere circuit breaker.

The dryer: 5000 watts divided by 240 volts equals 21 amperes. Therefore, you would use No. 10 wire with a 30-ampere circuit.

Laundry circuit: 1500 watts.

(c) Minimum size of service.

General lighting	7500 volt-amperes
Small appliance load	3000 volt-amperes
Laundry circuit	1500 volt-amperes
	12,000 volt-amperes
3000 volt-amperes @ 100 percent	3000 volt-amperes
9000 volt-amperes @ 35 percent	3150 volt-amperes
Net computed (without range and dryer)	6150 volt-amperes
Range load	8000 volt-amperes
Dryer load	5000 volt-amperes
	13,000 volt-amperes
Net computed (with range and dryer)	19,150 volt-amperes

Loads of over 10 kW must have a minimum of 100-ampere service; therefore, the minimum service will be 100 amperes. You can use No. 4 copper THW conductors with a 100-ampere main disconnect (see NEC, *Note 3* to *Tables 310.16* through *310.31*).

5-62 **What are the demand factors for nondwelling receptacle loads?**
See NEC, *Table 220.13*.

5-63 **How are motor loads computed?**
In accordance with *Sections 430.24, 430.25,* and *430.26* (see NEC, *Section 220.14*).

5-64 **How are fixed electric space heating loads computed?**
At 100 percent of the total connected load, but in no case is a feeder load current rating to be less than the rating of the largest branch-circuit supplied (see NEC, *Section 220.15*).

5-65 **What percentage is permitted to be applied as a demand factor to the nameplate rating load for four or more appliances fastened in place and served by the same feeder in a one-family, two-family, or multifamily dwelling?**
Seventy-five percent for all appliances except electric ranges, clothes dryers, space heating equipment, or air-conditioning equipment (see NEC, *Section 220.17*).

5-66 **The demand factor for five household electric clothes dryers is calculated at what percentage?**
80 percent (see NEC, *Table 220.18*).

5-67 **Where two or more single-phase ranges are supplied by a 3-phase, 4-wire feeder, how is the total load computed?**
On the basis of twice the maximum number connected between any two phases (see NEC, *Section 220.19*).

5-68 **What is the maximum demand for two household electric ranges, wall-mounted ovens, counter-mounted cooking units, and other household cooking appliances rated at over 1¾ kW?**
11 kW for ratings not over 12 kW (see NEC, *Section 220.19*).

5-69 **What computations are required for commercial cooking equipment?**
See NEC, *Table 220.20*.

5-70 **When computing the total load of a feeder for an electric heating system and an air-conditioning system that won't be used simultaneously, must both loads be added?**
No. The smaller of the two can be omitted (see NEC, *Section 220.21*).

5-71 A store building is to be wired for general illumination and show-window illumination. The store is 40 feet by 75 feet, with 30 linear feet of show window. With a density of illumination of 3 watts per square foot for the store and 200 volt-amperes per linear foot for the show-window, calculate (a) the general store load, (b) the minimum number of branch circuits required and sizes of wire, and (c) the minimum size of feeders (or service conductors) required.

(a) General lighting load. 3000 square feet at 3 volt-amperes per square foot equals 9000 volt-amperes. However, this load will be required most of the time, so you must multiply 1.25 by 9000, which yields 11,250 volt-amperes.

Show window: 30 linear feet at 200 volt-amperes per foot equals 6000 volt-amperes. Therefore, the general store load is 11,250 volt-amperes.

(b) Minimum number of branch circuits and wire sizes. 11,250 volt-amperes divided by 240 volts equals 47 amperes. For three-wire service, this current will require four 15-ampere circuits using No. 14 wire (minimum) or three 20-ampere circuits using No. 12 wire (minimum).

Show window: 6000 volt-amperes divided by 240 volts equals 25 amperes. Therefore, two 15-ampere circuits with No. 14 wire or two 20-ampere circuits with No. 12 wire will be required.

(c) Minimum size service conductors required. The ampere load for three-wire service would be 47 amperes plus 25 amperes, which equals 72 amperes. Therefore, a 100-amp service would be used with No. 3 THW conductors in 1¼-inch conduit or No. 3 THHN conductors in 1-inch conduit.

If the service is 120 volts instead of 120/240 volts (which is highly improbable because the utility serving would require the 120/240 volts), then you would have to use 120 volts when finding the current. These currents would then be doubled; entrance-wire capacities would double as well as the main disconnect on the service entrance.

5-72 Determine the general lighting and appliance load requirements for a multifamily dwelling (apartment house). There are 40 apartments, each with a total of 800 square feet. There are two banks of meters of 20 each, and individual subfeeders to each apartment. Twenty apartments have electric ranges; these apartments (with ranges) are evenly divided, 10 on each meter bank, and the ranges are 9 kW each. The service is 120/240 volts. Make complete calculations of what will be required, from the service entrance on. Computed load for each apartment (see NEC, *Article 220*).

General lighting load:

800 square feet @ 3 volt-amperes	
per square foot	2400 volt-amperes
Small-appliance load	3000 volt-amperes
Electric range	3000 volt-amperes

Minimum number of branch circuits required for each apartment (see NEC, *Section 220.4*).

2400 volt-amperes divided by 120 volts equals 20 amperes. This current will require two 15-ampere circuits using No. 14 wire or two 20-ampere circuits using No. 12 wire.

Small-appliance load: Two 20-ampere circuits using No. 12 wire (scc NEC, *Section 220.4(B)*).

Range circuit: 8000 watts divided by 240 equals 33 amperes. A circuit of two No. 8 wires or one No. 10 wire is required (see NEC, *Section 210.19(B)*).

Minimum size subfeeder required for each apartment (see NEC, *Section 215.2*).

Computed load:

General lighting load	2400 volt-amperes
Small-appliance load (two 20-ampere circuits)	3000 volt-amperes
Total computed load (without ranges)	5400 volt-amperes
Application of demand factor:	
3000 watts @ 100 percent	3000 volt-amperes
2400 watts @ 35 percent	840 volt-amperes
Net computed load (without ranges)	3840 volt-amperes
Range load	8000 volt-amperes
Net computed load (with ranges)	11,840 volt-amperes

For 120/240 volt, 3-wire system (without ranges):

Net computed load: 3840 volt-amperes divided by 240 volts equals 16 amperes.

Minimum feeder size could be No. 10 wire with a two-pole, 30-ampere circuit breaker.

For 120/240 volt, 3-wire system (with ranges):

> Net computed load: 11,840 volt-amperes divided by 240 volts equals 49 amperes.
>
> Minimum feeder size would be No. 6 wire with two 60-ampere fuses or No. 4 wire with a two-pole, 70-ampere circuit breaker.

Neutral subfeeder:

Lighting and small-appliance load	3840 volt-amperes
Range load (8000 watts @ 70 percent)	
(see NEC, *Section 220-22*)	5600 volt-amperes
Net computed load (neutral)	9440 volt-amperes

9400 watts divided by 240 volts equals 41 amperes. Size of neutral subfeeder would be No. 6 wire. There would also be two main disconnects needed ahead of the meters.

Minimum size feeders required from service equipment to meter bank (for 20 apartments—10 with ranges).

> Total computed load: Lighting and small-appliance load—20 multiplied by 5400 volt-amperes equals 108,000 volt-amperes.

Application of demand factor:

3000 volt-amperes @ 100 percent	3000 volt-amperes
105,000 volt-amperes @ 35 percent	36,750 volt-amperes
Net computed lighting and small- appliance load	39,750 volt-amperes
Range load (10 ranges, less than 12 kW)	25,000 volt-amperes
Net computed load (with ranges)	64,750 volt-amperes

For 120/240 volt, 3-wire system: Net computed load—64,750 divided by 240 equals 270 amperes. Size of each ungrounded feeder to each meter bank would be 500,000 CM.

Neutral feeder:

Lighting and small-appliance load	39,750 volt-amperes
Range load (25,000 watts @ 70 percent)	17,500 volt-amperes
Computed load (neutral)	57,250 volt-amperes

57,250 volt-amperes divided by 240 volt equals 239 amperes.

Further demand factor:

200 amperes @ 100 percent	200 amperes
39 amperes @ 70 percent	27 amperes
Net computed load (neutral)	227 amperes

Minimum size main feeder (service conductors) required (for 40 apartments—20 with ranges).

Total computed load: Lighting and small-appliance load—40 multiplied by 5400 volt-amperes equals 216,000 volt-amperes.

Application of demand factor:

3000 volt-amperes @ 100 percent	3000 volt-amperes
117,000 volt-amperes @ 35 percent	40,950 volt-amperes
96,000 volt-amperes @ 25 percent	24,000 volt-amperes
Net computed lighting and small-appliance load	67,950 volt-amperes
Range load (20 ranges, less than 12 kW)	35,000 volt-amperes
Net computed load	102,950 volt-amperes

For 120/240 volt, 3-wire system:

Net computed load—102,950 volt-amperes divided by 240 volts equals 429 amperes.

Size of each ungrounded main feeder would be 1,000,000 CM.

Neutral feeder:

Lighting and small appliance load	67,950 volt-amperes
Range load (35,000 volt-amperes @ 70 percent)	24,500 volt-amperes
Computed load (neutral)	92,450 volt amperes

92,450 volt-amperes divided by 240 volts equals 385 amperes.

Further demand factor:

200 amperes @ 100 percent	200 amperes
185 amperes @ 70 percent	130 amperes
Net computed load (neutral)	330 amperes

Size of neutral main feeder would be 6,000,000 CM.

5-73 There is to be a current of 100 amperes per phase on a 4-wire 120/208-volt wye system. What size neutral would you need? The phase wires are No. 2 THHN.

Since in this case neutrals are not allowed to be derated and must be the same size as the phase conductors, you would need to use a No. 2 THHN conductor.

5-74 Is there a demand factor for feeders and service-entrance conductors for multifamily dwellings?

Yes (see NEC, *Table 220.32*).

Chapter 6

Transformer Principles and Connections

6-1   What is induction?
The process by which one conductor produces, or induces, a voltage in another conductor, even though there is no mechanical coupling between the two conductors.

6-2   What factors affect the amount of induced electromotive force (emf) in a transformer?
The strength of the magnetic field, the speed at which the conductors are cut by the magnetic field, and the number of turns of wire being cut by the magnetic field.

6-3   What is inductance?
The property of a coil in a circuit to oppose any change of existing current flow.

6-4   What is self-inductance?
The inducing of an emf within the circuit itself, caused by any change of current within that circuit. This induced emf is always in a direction opposite to the applied emf, thus causing opposition to any change in current within the circuit itself.

6-5   What is mutual inductance?
The linkage of flux between two coils or conductors, caused by the current flowing within one or both of the coils or conductors.

6-6   Draw a diagram of two coils, such as the coils of a transformer winding, and indicate the self-inductance and the mutual inductance.
Self-inductance is produced within the primary coil, and mutual inductance exists between the two transformer coils, as shown in Figure 6-1.

6-7   Name several methods by which an emf can be generated.
By conductors being cut by a magnetic field (as in generators), by chemical reactions (as in batteries), by heat (as in thermocouples), by crystal vibration (as in piezoelectricity), and by friction (as in static electricity).

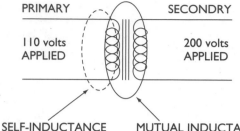

SELF-INDUCTANCE   MUTUAL INDUCTANCE

Figure 6-1 Self-inductance and mutual inductance in the coils of a transformer.

6-8 What is direct current (dc)?
Current that flows in one direction only.

6-9 What is alternating current (ac)?
Current that continually reverses its direction of flow.

6-10 What is pulsating direct current?
An unidirectional current that changes its value at regular or irregular intervals.

6-11 What is a cycle?
One complete alternation, or reversal, of alternating current. The wave rises from zero to maximum in one direction, falls back to zero, then rises to maximum in the opposite direction, and finally falls back to zero again.

6-12 What always surrounds a conductor when a current flows through it?
A magnetic field.

6-13 What is the phase relation between the three phases of a three-phase circuit?
They are 120 electrical degrees apart.

6-14 Draw sine waves for three-phase voltage; show polarity, time, and phase angle (in degrees).
See Figure 6-2.

6-15 What is the phase relation between phases in two-phase circuit?
They are 90 electrical degrees apart.

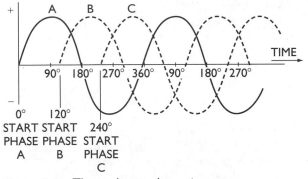

Figure 6-2   Three-phase voltage sine waves.

6-16 What is a transformer?

A device that transforms electrical energy from one or more circuits to one or more other circuits at the same frequency but usually at a different voltage and current. It consists of a core of soft-iron laminations surrounded by coils of copper-insulated wire.

6-17 There are two basic types of transformers. What are they?

The isolation type, in which the two windings are physically isolated and electrically insulated from each other, and the autotransformer type, in which there is only one coil with a tap or taps taken off it to secure other voltages (the primary is part of the secondary and the secondary is part of the primary).

6-18 What is an oil-immersed transformer?

The core and coils are immersed in a high-grade mineral oil, which has high dielectric qualities.

6-19 Why is oil used in a transformer?

To increase the dielectric strength of the insulation, to keep down the possibility of arcing between coils, and to dissipate heat to the outer case so that the transformer can carry heavier loads without excessive overheating.

6-20 What is an air-core transformer?

A transformer that does not contain oil or other dielectric compositions but is insulated entirely by the winding insulations and air.

6-21 What are eddy currents?
Circulating currents induced in conductive materials (usually the iron cores of transformers or coils) by varying magnetic fields.

6-22 Are eddy currents objectionable?
Yes, they represent a loss in energy and also cause overheating.

6-23 What means can be taken to keep eddy currents at a minimum?
The iron used in the core of an alternation-current transformer is laminated, or made up of thin sheets or strips of iron, so that eddy currents will circulate only in limited areas.

6-24 What is hysteresis?
When iron is subjected to a varying magnetic field, the magnetism lags the magnetizing force due to the fact that iron has reluctance, or resistance, to changes in magnetic densities.

6-25 Is hysteresis objectionable?
Yes, it is a loss and affects the efficiency of transformers.

6-26 Are transformers normally considered to be efficient devices?
Yes, they have one of the highest efficiencies of any electrical device.

6-27 What factors constitute the major losses produced in transformers?
Power loss of the copper (I^2R losses), eddy currents, and hysteresis losses.

6-28 Is there a definite relationship between the number of turns and voltages in transformers?
Yes, the voltage varies in exact proportion to the number of turns connected in series in each winding.

6-29 Give an illustration of the relationship between the voltages and the turns ratio in a transformer.
If the high-voltage winding of a transformer has 1000 turns and a potential of 2400 volts is applied across it, the low-voltage winding of 100 turns will have 240 volts induced across it; this is illustrated in Figure 6-3.

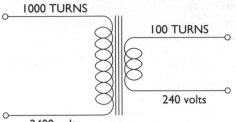

Figure 6-3 Relationship between voltages and the
turns ratio in a transformer.

6-30 What is the difference between the primary and the secondary of a transformer?

The primary of the transformer is the input side of the transformer and the secondary is the output side of the transformer. On a step-down transformer, the high-voltage side is the primary and the low-voltage side is the secondary; on a step-up transformer, the opposite is true (Figure 6-4).

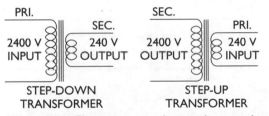

Figure 6-4 The primary and secondary windings of a
step-down and a step-up transformer.

6-31 Ordinarily, what is the phase relationship between the primary and secondary voltages of a transformer?

They are 180° out of phase.

6-32 Is it possible to have the primary and secondary of a transformer in phase?

Yes, by changing the connections on one side of the transformer.

6-33 How are the leads of a transformer marked, according to ANSI (American National Standards Institute)?

The high side of the transformer is marked H_1, H_2, etc. The low side of the transformer is marked X_1, X_2, etc.

6-34 What is the purpose of the markings on transformer leads?
They are there for standardization, so that transformer polarities are recognizable for any type of use.

6-35 Draw a diagram of a transformer with additive polarity, using ANSI markings.
See Figure 6-5.

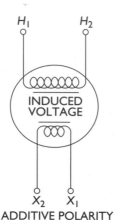

ADDITIVE POLARITY

Figure 6-5   A transformer with additive polarity.

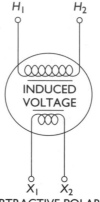

SUBTRACTIVE POLARITY

Figure 6-6   A transformer with subtractive polarity.

6-36 Draw a diagram of a transformer with subtractive polarity, using ANSI markings.
See Figure 6-6.

6-37 If a transformer is not marked, how could you test it for polarity?
Connect the transformer as shown in Figure 6-7. If it has subtractive polarity, V will be less than the voltage of the power source; if it has additive polarity, V will be greater than the voltage of the power source.

6-38 What is a split-coil transformer?
A transformer that has the coils on the low or high side in separate windings so that they can be connected in series or parallel for higher or lower voltages, as desired.

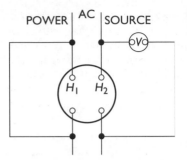

POWER | AC | SOURCE

Figure 6-7 Testing a transformer for polarity.

6-39 Draw a diagram of a split-coil transformer with the low side having split coils for dual voltages; draw an additive polarity transformer, and mark the terminals with ANSI markings. Show the voltages that you use.

See Figure 6-8.

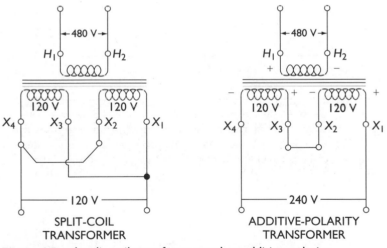

SPLIT-COIL TRANSFORMER

ADDITIVE-POLARITY TRANSFORMER

Figure 6-8 A split-coil transformer and an additive-polarity transformer.

6-40 Draw a diagram of an autotransformer.

See Figure 6-9.

6-41 Where may autotransformers be used?

(a) Where the system being supplied contains an identified grounded conductor that is solidly connected to a similar identified grounded conductor of the system supplying the autotransformer (see NEC, *Sections 210.9* and *450.4*); (b) where an autotransformer is used for starting or controlling an induction motor (see NEC, *Section 430.82(B)*); (c) where an autotransformer is used as a dimmer, such as in theaters (see NEC, *Section 520.25(C)*); (d) as part of a ballast for supplying lighting units (see NEC, *Section 410.78*). For voltage bucking and boosting, see NEC, *Section 210.9 exception 1.*

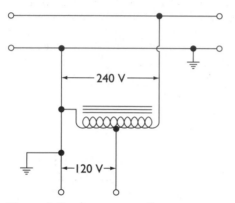

Figure 6-9 An autotransformer.

6-42 What is the relationship between the current and voltage in the high side of a transformer and the current and voltage in the low side of a transformer? Draw a diagram showing this relationship.

With respect to the turns ratio, the current in one side of a transformer is inversely proportional to the current in the other side, whereas the voltage across one side of a transformer is directly proportional to the voltage across the other side. These are illustrated in Figure 6-10.

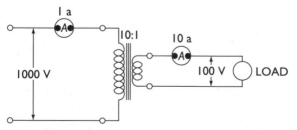

Figure 6-10 Current-voltage relationship between the high side and the low side of a transformer.

6-43 When an induction coil is connected in a dc circuit, as in Figure 6-11, what happens when the switch is closed? When the switch is opened?

When the switch is closed, the current slowly rises to a maximum (point *A* in this example). The retarding of current flow is due to self-inductance. After reaching the maximum at point *A*, the current will remain constant until the switch is opened (point *B*). When the switch is opened, the flux around the coil collapses, thereby causing an opposition to the current discharge; however, this discharge-time collapse is extremely short when compared to the charging time. The discharge causes a high voltage to be applied across the switch, which tends to sustain an arc; this voltage often reaches large values. The principles of this type of circuit have many applications, such as ignition coils and flyback transformers.

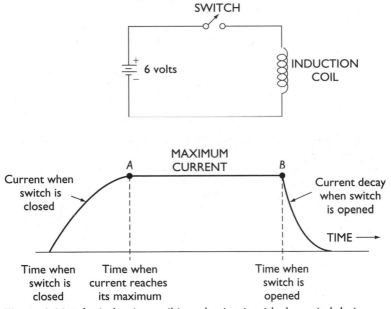

Figure 6-11   An induction coil in a dc circuit with the switch being opened and closed.

6-44 Draw a schematic diagram of the high-side windings of three single-phase transformers connected in a delta arrangement. Show the ANSI markings.

Note that in a delta arrangement, as shown in Figure 6-12, H_1 is connected to H_2 of the next transformer, and so on.

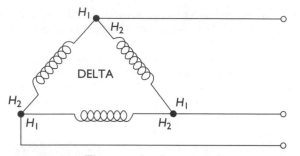

Figure 6-12   Three single-phase transformer windings connected in a delta arrangement.

6-45 Draw a schematic diagram of the high-side windings of three single-phase transformers connected in a wye (Y) arrangement. Show the ANSI markings.

Note that in a wye arrangement, as shown in Figure 6-13, all of the H_2s are connected in common, and the H_1s each supply one phase wire.

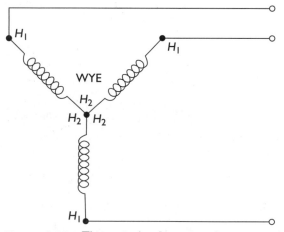

Figure 6-13   Three single-phase transformer windings connected in a wye arrangement.

6-46 If the polarity of one transformer is reversed on a delta bank of single-phase transformers that are connected for three-phase operation, what would be the result?

Instead of zero voltage on the tie-in point, there would be a voltage that is twice the proper value.

6-47 Show with diagrams how you would test to be certain that the polarities on a delta bank of three single-phase transformers are correct.
See Figure 6-14.

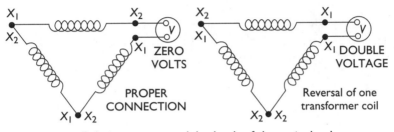

Figure 6-14 Polarity tests on a delta bank of three single-phase transformer windings.

6-48 In a bank of three single-phase transformers that are connected in a delta, each transformer delivers 240 volts at 10 amperes. What are the line voltages and line currents?
The line voltages are each equal to 240 volts; however, the line current in each phase would be the current of each transformer multiplied by 1.732 (the square root of 3), or 17.32 amperes.

6-49 Draw a schematic diagram showing all the currents and voltages on a bank of three single-phase transformers that are connected in a delta arrangement. Assume a voltage across each transformer of 240 volts and a current through each transformer of 10 amperes.
See Figure 6-15.

6-50 Draw a schematic diagram for a bank of three single-phase transformers that are connected in a wye. Show all voltages and currents; assume a voltage across each transformer of 120 volts and a current in each winding of 10 amperes.
See Figure 6-16.

6-51 Draw a bank of three single-phase transformers that are connected in a delta-delta bank with one side connected to 2400 volts, three-phase, and with the other side delivering 240 volts, three-phase. Show voltages and ANSI markings on all transformers.
See Figure 6-17.

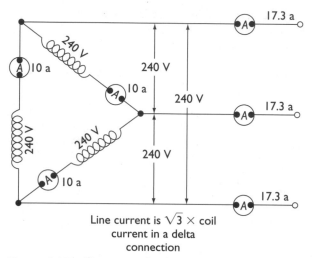

Line current is $\sqrt{3}$ × coil current in a delta connection

Figure 6-15 Current and voltage values on a delta bank of three single-phase transformer windings. A voltage of 240 volts and a current of 10 amperes are assumed.

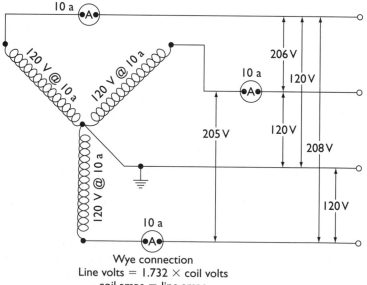

Wye connection
Line volts = 1.732 × coil volts
coil amps = line amps

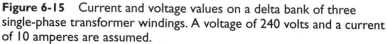

-16 Current and voltage values on a wye bank of three ase transformer windings. A voltage of 120 volts and a current peres are assumed.

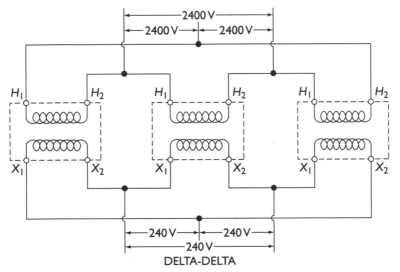

Figure 6-17   Three single-phase transformers connected in a delta-delta bank; the high side is connected to 2400 volts, three-phase, and the low side delivers 240 volts, three phase.

6-52 Is it possible to connect two single-phase transformers to secure a three-phase output from a three-phase input?

Yes, they would have to be connected in an open delta.

6-53 If you have a bank of three single-phase transformers that are connected in a closed delta arrangement, and one transformer burns up, how would you continue operation on the remaining two transformers?

By merely disconnecting the leads to the disabled transformer.

6-54 When you use a bank of two single-phase transformers in an open delta arrangement, do they supply their full output rating?

No. Each transformer is only capable of supplying 86.6 percent of its output rating.

6-55 If you have a bank of three single-phase transformers, each with a 10-kVA rating, that are connected in a closed delta arrangement, you would have a capacity of 30 kVA. If one transformer is taken out of the bank, what would be the output capacity of the remaining 10-kVA transformers?

Each transformer would deliver 8.66 kVA, and you would have a bank capacity of 17.32 kVA.

6-56 Draw a schematic diagram of two transformers that are connected in an open delta arrangement. Show transformer voltages and the three-phase voltages.
    See Figure 6-18.

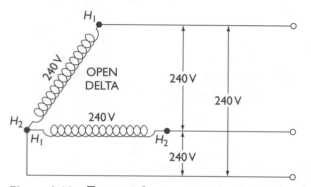

Figure 6-18   Two transformers in an open delta arrangement.

6-57 Draw a schematic diagram of three transformers that are connected in a delta arrangement on both sides, fed from a 2400-volt source on the high side, and connected for a 240-volt, three-phase, and a 120/240-volt, single-phase output on the low side. Show all voltages.
    See Figure 6-19.

6-58 Draw a schematic diagram of three single-phase transformers that are connected in a wye-wye arrangement. Show the neutral on both high and low sides.
    See Figure 6-20.

6-59 Draw a schematic diagram of three single-phase transformers that are connected in a wye arrangement on the high side and a delta arrangement on the low side.
    See Figure 6-21.

6-60 What are instrument transformers?
    In the measurement of current, voltage, or kilowatt-hours on systems with high voltage or high current, it is necessary to use a device known as an instrument transformer, which reproduces in its secondary circuit the primary current or voltage while preserving the phase relationship to measure or record at lower voltages or lower amperages, and then to use a constant to multiply the readings to obtain the actual values of voltage or current. Current transformers (CTs) are used to measure the current, and potential transformers (PTs) are used to register the potential.

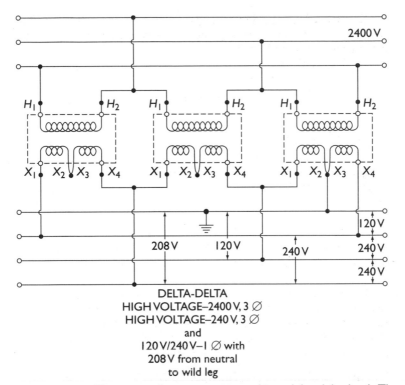

DELTA-DELTA
HIGH VOLTAGE–2400 V, 3 ∅
HIGH VOLTAGE–240 V, 3 ∅
and
120 V/240 V–1 ∅ with
208 V from neutral
to wild leg

Figure 6-19   Three transformers connected in a delta-delta bank. The high side is connected to 2400 volts, three-phase, and the low side delivers 240 volts, three-phase, and 120/240 volts, single-phase.

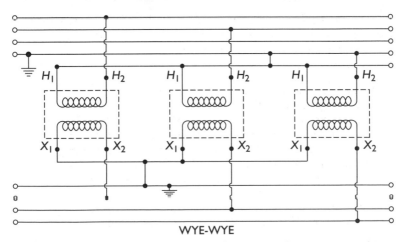

WYE-WYE

Figure 6-20   Three single-phase transformer windings connected in a wye-wye arrangement.

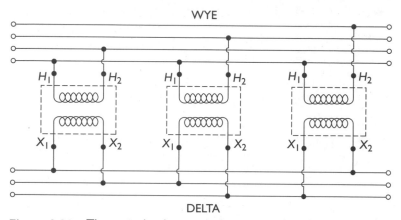

WYE

DELTA

Figure 6-21   Three single-phase transformer windings connected in a wye arrangement on the high side and a delta arrangement on the low side.

6-61 Describe a potential transformer.

A potential transformer is built like the ordinary isolation transformer, except that extra precautions are taken to ensure that the winding ratios are exact. Also, the primary winding is connected in parallel with the circuit to be measured.

6-62 Describe a current transformer.

A current transformer has a primary of a few turns of heavy conductor capable of carrying the total current, and the secondary consists of a number of turns of smaller wire. The primary winding is connected in series with the circuit carrying the current that is to be measured.

6-63 How are current transformers rated?

They are rated at 50 to 5, 100 to 5, etc. The first number is the total current that the transformer is supposed to handle, and the second figure is the current on the secondary when the full-load current is flowing through the primary. For example, a 50-to-5 rating would have a multiplier of 10 ($K = 10$).

6-64 What precautions must be taken when working with current transformers? Why?

The secondary must never be opened when the primary circuit is energized. If it is necessary to disconnect an instrument while the circuit is energized, the secondary must be short-circuited. If the secondary is opened while the circuit is energized, the potential on the secondary might reach dangerously high values. By short-circuiting the secondary, damage is avoided and the voltage on the secondary is kept within safe limits.

6-65 Draw a schematic diagram of a current transformer.

As shown in Figure 6-22, the primary consists of a single conductor; it may be a single conductor or only a few turns.

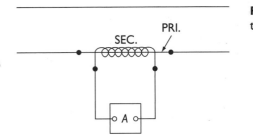

Figure 6-22 A current transformer.

6-66 What is a booster transformer?

A transformer arrangement that is often used toward the end of a power line in order to raise the voltage to its desired value. These are often called "Buck-boost" transformers.

6-67 May an ordinary transformer be used as a booster transformer?
Yes.

6-68 When connecting an ordinary transformer as a booster transformer, what important factors must be considered?

The high side of the transformer must be able to handle the approximate voltage of the line; the low side must have a voltage of approximately the value by which you wish to boost the line voltage and must also have a current capacity that is sufficient to carry the line current.

6-69 What special precaution must be taken when using a booster transformer?

There must be no fusing in the high side, or primary. Because the booster transformer is similar to a current transformer, an extremely high voltage could be built up on the secondary side if the fuse should blow.

6-70 Draw a schematic diagram of an ordinary transformer that is connected as a booster transformer.
See Figure 6-23.

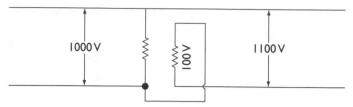

Figure 6-23   A voltage-booster transformer.

6-71 What is an induction regulator?

This device is similar to a booster transformer. It has a primary and a secondary winding, which are wound on separate cores. The primary can be moved in either direction; this is usually done by an electric motor. In turning, the primary bucks or boosts the line voltage, as required. The amount of bucking or boosting is anticipated by the current being drawn by the line.

6-72 Draw a schematic diagram showing how an induction regulator is connected into the line.

See Figure 6-24.

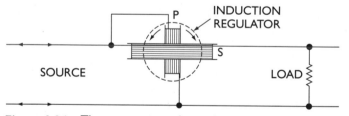

Figure 6-24   The connection of an induction regulator to the line.

6-73 What is a three-phase transformer?

A transformer that is the equivalent of three single-phase transformers, which are all wound on one core and enclosed within one common case.

6-74 When connecting transformers in parallel, what factors must be taken into consideration?

Their electrical characteristics, such as voltage ratio, impedance percentage, and voltage regulation.

6-75 If transformers with different electrical characteristics are connected in parallel, what will happen?

They won't distribute the load equally; one transformer will tend to assume more of the load than the other. This leads to overheating and, in severe cases, the destruction of the transformer(s).

Chapter 7

Wiring Design and Protection

7-1 What are the minimum requirements for service-drop conductors?

They must be of sufficient size to carry the load that is required of them, but they musn't be smaller than No. 8 copper wire or No. 6 aluminum, except under limited load conditions where they may not be smaller than No. 12 hard-drawn copper (see NEC, *Section 230.23*).

7-2 What is the minimum clearance for service drops over buildings?

They shall have a minimum clearance of 8 feet (see NEC, *Section 230.24(A)*). However, if the voltage does not exceed 300 volts between conductors and the roof has a slope of not less than 4 inches in 12 inches, the clearance may be a minimum of 3 feet (see NEC, *Section 230.24(A)*, Exception No. 2).

7-3 What is the minimum height of point of attachment of service drops?

10 feet, provided the clearance in the NEC, *Section 230.24(B)* is met (see also NEC, *Section 230.26*).

7-4 What is the minimum clearance of service drops over commercial areas, parking lots, agricultural areas, and other areas subject to truck traffic?

18 feet (see NEC, *Section 230.24(B)*).

7-5 What is the minimum clearance of service drops over sidewalks?

12 feet (see NEC, *Section 230.24(B)*).

7-6 What is the minimum clearance of service drops over driveways, alleys, and public roads?

18 feet (see NEC, *Section 230.24(B)*).

7-7 What is the minimum clearance of service drops over residential driveways?

12 feet (see NEC, *Section 230.24(B)*).

7-8   Can a bare neutral conductor be buried in the ground in an underground service?

No; it must be insulated, unless it is in a duct or conduit (see NEC, *Section 230.30*). There is an exception: bare copper for direct burial where bare copper is judged to be suitable for the soil conditions.

7-9   Where underground service conductors are carried up a pole, to what minimum height must they be given mechanical protection?

8 feet (see NEC, *Section 300.5(C)(1)*).

7-10   Is sealing required where underground ducts or conduits enter buildings?

Yes, to prevent the entrance of gases or moisture into the building. Spare or unused conduits and ducts must also be sealed (see NEC, *Section 230.8*).

7-11   What is the minimum size of service-entrance conductor normally allowed?

Any approved conductor having an ampacity equal to the sum of the noncontinuous loads, plus 125 percent of continuous loads. However, if the service entrance conductors terminate in an overcurrent device and assembly listed for operation at 100 percent of its rating, the 125 percent multiplier need not be used (see NEC, *Section 230.42(A)*).

7-12   What is the minimum size service for a single-family dwelling with an initial load of 10 kVA or more?

The service must be a minimum of 100 amperes for a 3-wire service (see NEC, *Section 230.79(C)*).

7-13   A residence has more than six 2-wire branch circuits. What is the minimum size of service required for this residence?

The service entrance conductors shall have an ampacity of not less than 100 amperes for a 3-wire service (see NEC, *Section 230.79(C)*).

7-14   Are splices permitted in service-entrance conductors?

Yes. They may be spliced with clamped or bolted connections; or buried splices may be made with a listed underground splice kit (see NEC, *Section 230.46, 110.4, 300.5(E), 300.13*, and *300.15*).

7-15 Are conductors other than service-entrance conductors permitted in the same raceways or cables?

No, except for grounding conductors and load management control conductors (see NEC, *Section 230.7*, with exceptions).

7-16 What must be provided for the disconnection of service conductors from building conductors?

Some means must be provided for disconnecting all of the building conductors from the service-entrance conductors, and this means must be located in a readily accessible point nearest the point of entrance of the service conductors, on either the outside or inside of the building, whichever is most convenient (see NEC, *Section 230.70*). Also see definition in *Article 100* for Service Equipment.

7-17 What must the service disconnecting means consist of?

It may consist of not more than six switches or six circuit breakers in a common enclosure or in a group of separate enclosures, provided that they are grouped together. Two or three single-pole switches or circuit breakers may be installed on multiwire circuits and counted as one, provided they have handle ties or handles, so that not more than six operations of the hand are required to disconnect all circuits (see NEC, *Section 230.71(A)*).

7-18 What may be connected ahead of the service entrance disconnecting switch?

Service fuses, meters, high-impedance shunt circuits (such as potential coils of meters), supply conductors for time switches, surge-protection capacitors, instrument transformers, lightning arresters, solar electric systems, interconnected electric power production sources, and circuits for emergency systems (such as fire-pump equipment, etc.) (see NEC, *Section 230.82*).

7-19 In multiple-occupancy dwellings, must each occupant have access to his disconnecting means?

Yes, except if the disconnects are under constant supervision and maintenance; in this case, they can be accessible to the building management only (see NEC, *Section 230.72(C)* and Exception).

7-20 Is it permissible to install overcurrent-protective devices in the grounded service conductor?

Overcurrent-protective devices may not be installed in the grounded service conductor unless a circuit breaker is used that opens all conductors of the circuit simultaneously (see NEC, *Section 230.9(B)*).

7-21 What is the minimum size of service-entrance conductors operating at over 600 volts?
They musn't be smaller than No. 6 wire, unless in cable, where they may be a minimum of No. 8 (see NEC, *Section 230.202(A)*).

7-22 What is the maximum circuit protection allowed with flexible cords, sizes No. 16 or No. 18, and tinsel cord?
30 and 20 amperes (see NEC, *Section 240.5(B)(1)*).

7-23 What is the maximum overcurrent protection allowed on cords of 20 amperes capacity?
50 amperes (see NEC, *Section 240.5(B)(1)*).

7-24 Give the standard ratings (in amperes) for fuses and circuit breakers.
Standard ratings are 15, 20, 25, 30, 35, 40, 45, 50, 60, 70, 80, 90, 100, 110, 125, 150, 175, 200, 225, 250, 300, 350, 400, 450, 500, 600, 700, 800, 1000, 1200, 1600, 2000, 2500, 3000, 4000, 5000, 6000 (see NEC, *Section 240.6*).

7-25 Where can an overcurrent device be used in a grounded conductor?
No overcurrent device shall be placed in any permanently grounded conductor, except as follows:
Where the overcurrent device simultaneously opens all conductors of the circuit (see *Section 240.22 of the NEC*). A fuse shall also be inserted in the grounded conductor when the supply is 3-wire, 3-phase ac, one conductor grounded (see *Section 430.36 of the NEC*).

7-26 May fuses be arranged in parallel?
No, except as factory assembled in parallel and approved as a unit (see NEC, *Section 240.8*).

7-27 May breakers be paralleled?
No, only circuit breakers assembled in parallel that are tested and approved as a single unit (see *Section 240.8 of the NEC*).

7-28 Are the secondary conductors of transformer feeder taps required to have overcurrent protection?

Yes, unless they meet the requirements of *Section 240.21(C) (1)* through *(6)*.

7-29 In what position must a knife switch with fuses be mounted? Why?

In a vertical position, so that when the switch is opened, the blades won't close by gravity; the line side must be connected so that when the switch is opened, the fuses will be deenergized (see *Section 404.6(A)* of the NEC).

7-30 When overcurrent-protective device enclosures are mounted in a damp location, what precautions must be taken?

The enclosures must be of an identified type for the location and must be mounted with at least a ¼-inch air space between the wall or supporting surface (see NEC, *Sections 312.2(A)* and *240.32*).

7-31 Where may Edison-base plug fuses be used?

Plug fuses of the Edison-base type shall be used only for replacement in existing installations where there is no evidence of overfusing or tampering (see NEC, *Section 240.51(B)*).

7-32 What is the maximum voltage rating of plug fuses?

125 volts (see NEC, *Section 240.50(A)*).

7-33 What precautions must be taken if plug fuses are used on new installations?

Fuses, fuseholders, and adapters must be so designed that other Type S fuses may not be used (see NEC, *Section 240.53(B)*).

7-34 What are the size classifications of Type S fuses and adapters?

Not exceeding 125 volts at 0–15 amperes, 16–20 amperes, and 21–30 amperes (see NEC, *Section 240.53(A)*).

7-35 What is a Type S fuse?

A dual element fuse with special threads so that fuses larger than what the circuit was designed for may not be used. They are designed to make tampering or bridging difficult (see NEC, *Section 240.54*).

7-36 What are the overcurrent protection requirements for listed extension cord sets?

See NEC, *Section 240.5(B)*.

7-37 Why are systems and circuits grounded?
To limit the excess voltage to ground, which might occur from lightning or exposure to other higher-voltage sources (see NEC, *Section 250.2*).

7-38 Is it necessary to ground one wire on a two-wire dc system with not more than 300 volts between conductors?
Yes, in practically every case, although there are a few exceptions (see NEC, *Section 250.162*).

7-39 When should ac systems be grounded?
Ac circuits and systems must be grounded as provided for in *Section 250.20* of the NEC.

7-40 When is the next higher overcurrent protective device permitted to protect a conductor where a standard fuse or circuit breaker is not available?
In general up to 800 amperes (see NEC, *Section 240.3*).

7-41 How are remote control circuits protected against overcurrents?
In accordance with *Article 725* (see NEC, *Section 240.3*).

7-42 What article in the NEC covers protection from overcurrent for industrial machinery?
Article 670 (see NEC, *Section 240.3*).

7-43 When do circuits of 50 volts or less to ground have to be grounded? When do they not have to be grounded?
Circuits of less than 50 volts need not be grounded unless they are supplied from systems exceeding 150 volts to ground, where they are supplied by transformers from ungrounded systems, or where they run overhead outside buildings (see NEC, *Section 250.20(A)*).

7-44 If a grounded conductor is not going to be used in a building, must this grounded conductor be run to each service?
Yes. The grounded conductor must be run to each service. It need not be taken farther than the service equipment if it won't be required in any of the circuits (see NEC, *Section 250.24*).

7-45 Is it necessary to ground each service on a grounded ac system?
Each individual service must have a ground, and this ground must be connected on the supply side of the service-disconnecting

means. If there is only one service from a transformer, there must be additional ground connection at the transformer or elsewhere in the transformer circuit for the installation to be approved (see NEC, *Section 250.24*).

7-46 What are the grounding requirements when two or more buildings are supplied from one service?

Where two or more buildings or structures are supplied from a common service, the grounded system in each building or structure is required to have a grounded electrode as described in NEC, *Section 250.52*, connected to the metal enclosure of the building disconnecting means and to the ac system grounded circuit conductor on the supply side of the building or structure disconnecting means (see *Section 250.32* of the NEC).

7-47 Is it required to ground a separately derived system, and if so, where?

Premises wiring systems that are required by *Section 250.20* to be grounded shall be grounded if the phase conductors are not physically connected to another supply system. They shall be grounded at the transformer, generator, or other source of supply, or at the switchboard on the supply side of the disconnecting means (see NEC, *Section 250.30*).

7-48 When must metal, noncurrent-carrying parts of fixed equipment that are likely to become energized be grounded?

When supplied by metal-clad wiring, when located in damp or wet places, when within reach of a person who can make contact with a grounded surface or object or within reach of a person standing on the ground, when in hazardous locations, or where any switches, enclosures, etc., are accessible to unqualified persons (see NEC, *Section 250.110*).

7-49 Must metal buildings be grounded?

If excessive metal in or on buildings may become energized and is subject to personal contact, adequate bonding and grounding will provide additional safety (see NEC, *Section 250.116*, FPN). However, this is a Fine Print Note to *Section 250.116* and not a requirement.

7-50 What is the exception to grounding the noncurrent-carrying parts of portable tools and appliances?

Portable tools and appliances may have double insulation. If they have, they must be so marked and won't require grounding (see NEC, *Section 250.114*, Exception).

7-51 What are the grounding requirements when lightning rods and conductors are present?

Lightning protection ground terminals must be bonded to the building's grounding system. In addition, metal enclosures with conductors should be kept at least 6 feet away from lightning conductors. Where this is not practical, they must be bonded together. However, this 6-foot rule is a Fine Print Note and not a requirement of the NEC. (See NEC, *Section 250.106*, with Exceptions.)

7-52 What is meant by effective grounding?

The path to ground must be permanent and continuous, must be capable of safely handling the currents that may be imposed on the ground, and must have sufficiently low impedance to limit the potential above ground to facilitate the opening of the overcurrent devices. The earth may not be used as the only equipment-grounding conductor (see NEC, *Section 250.2*).

7-53 What is a grounding electrode conductor used for?

The grounding conductor for circuits is used for grounding equipment, conduit, and other metal raceways, including service conduit, cable sheath, etc. (see NEC, *Section 250.24*).

7-54 What appliances may be grounded to the neutral conductor?

Electric ranges and electric clothes dryers, provided that the neutral is not smaller than No. 10 copper wire (see NEC, *Section 250.140*). However, this is not allowed for equipment that is fed from feeder panels, in mobile homes, or in recreational vehicles.

7-55 Three-wire SE cable with a bare neutral is sometimes used for connecting ranges and dryers. Is it permissible to use this type of cable when the branch circuit originates from a feeder panel?

No. The neutral must be insulated (see NEC, *Section 250.140(3)*).

7-56 What equipment, other than electric ranges and clothes dryers, may be grounded to the grounding conductor?

The grounding conductor on the supply side of the service-disconnecting means may ground the equipment, meter housing, etc. The load side of the disconnecting means cannot be used for grounding any equipment other than electric ranges and dryers (see NEC, *Section 250.142*).

7-57 How should continuity at service equipment be assured?
Threaded couplings and bosses, as well as threadless couplings and connections in rigid conduit or EMT (electrical metallic tubing), must be wrench-tight. Bonding jumpers must be used around concentric and eccentric knockouts and ordinary locknuts and bushings cannot be used for bonding (see NEC, *Section 250.92(B)*).

7-58 On flush-mounted grounded-type receptacles, is it necessary to bond the green grounding screw of the receptacle to the equipment ground?
Yes, with a few exceptions (see NEC, *Section 250.146*).

7-59 Can conduit serve as the equipment ground?
Except for a few cases, such as special precautions in hazardous locations, conduit, armored cable, metal raceways, etc., can serve as the equipment ground (see NEC, *Section 250.11862*).

7-60 What is required to assure electrical continuity of metal raceways and metal-sheathed cable used on voltages exceeding 250 volts?
The electrical continuity of metal raceways or metal-sheathed cable that contains any conductor other than service-entrance conductors of more than 250 volts to ground shall be assured by one of the methods specified in *Section 250.92(B)*, or by one of the following methods (refer to *Section 250.97* of the NEC):

(a) Threadless fittings, made up tight, with conduit or metal-clad cable.

(b) Two locknuts, one inside and one outside the boxes and cabinets.

(c) Fittings with shoulders that seat firmly against the enclosure, with a locknut on the inside of the enclosure, such as EMT fittings.

(d) Listed fittings identified for such use.

7-61 What is the preferred type of grounding electrode?
A metal underground water system where there are 10 feet or more of buried metal pipe, including well casings that are bonded to the system, is the preferred grounding electrode (see NEC, *Section 250.52*). A made electrode shall also be used in addition to the buried metal water pipe.

7-62 Is reinforcing bar in concrete permitted for a grounding electrode when buried metal water piping is not available?
Yes (see NEC, *Section 250.52(A)(3)*).

7-63 What is a "made" electrode?
A made grounding electrode may be a driven pipe, driven rod, buried plate, or other metal underground system approved for the purpose of grounding the equipment (see NEC, *Sections 250.50* and *250.52*). Note that the term "made electrode" is falling from use.

7-64 What are the requirements for plate electrodes?
They musn't have less than 2 square feet (.186 square meter) of surface exposed to the soil. Electrodes of iron or steel must be at least ¼ inch (6.4mm) thick; if made of nonferrous metal, it must be at least 0.06 inch (1.5mm) thick (see NEC, *Section 250.52(A)(6)*).

7-65 What are the requirements for pipe electrodes?
They must be at least ¾-inch trade size, and, if made of iron or steel, they must be galvanized or otherwise metal-coated to prevent corrosion; they must also be driven to a depth of at least 8 feet (see NEC, *Section 250.52(A)(5)(a)*).

7-66 What are the requirements for rod electrodes?
Electrodes of iron and steel must be at least ⅝ inch in diameter; if made of nonferrous material, they must be listed when ½ inch in diameter. Both types must be driven to a minimum depth of 8 feet (see NEC, *Section 250.52(A)(5)(b)*).

7-67 Describe the installation of made electrodes.
Unless rock bottom is encountered, they must be driven to a minimum depth of 8 feet, and below the permanent-moisture level. Where rock bottom is encountered at a depth of less than 4 feet, they must be buried 2½ feet deep horizontally in trenches or at oblique angles not more than 45 degrees (see NEC, *Section 250.53(G)*).

7-68 What should be the resistance of made electrodes?
If a single made electrode does not measure a resistance to ground of 25 ohms or less, it must be supplemented by another electrode that is not less than 6 feet away from the first electrode (see NEC, *Section 250.56*).

7-69 May grounding electrode conductors be spliced?

No, splices are not permitted; grounding electrode conductors must be one piece for their entire length, although there are some exceptions (see NEC, *Section 250.64(C)*).

7-70 What is the smallest size grounding electrode conductor permissible?

No. 4 wire or larger may be attached to buildings; No. 6 wire may be used if properly stapled to prevent physical damage; and No. 8 wire may be used if in conduit or armored cable for protection (see NEC, *Section 250.64(B)* and *Table 250.66*).

7-71 When a grounding electrode conductor is enclosed in a metal enclosure, how is the metal enclosure to be installed?

It must be electrically continuous from point of attachment to cabinet or enclosure to the point of attachment to the ground clamp or fitting. Metal enclosures that are not physically and electrically continuous must be bonded to the grounding electrode conductor at both ends of the metal enclosure (see NEC, *Section 250.64(E)*).

7-72 May aluminum be used for a grounding electrode conductor?

Yes, but it may not come in contact with masonry or the earth and cannot be run closer than 18 inches from the earth (see NEC, *Section 250.64(A)*).

7-73 If a water pipe is used as the grounding electrode, what precautions must be taken?

The grounding connection should be made at the point of water-service entrance. If this cannot be done and there is a water meter on the premises, the water meter must be bonded with a jumper of sufficient length, so that the water meter may be readily removed without disturbing the bonding. The cold-water piping must be used, and it should be checked to make certain that there are no insulated connections in the piping (see NEC, *Section 250.68(B)*).

7-74 Where nonmetallic water pipe serves an apartment building, what grounding procedures are recommended?

The ground must also be bonded to the interior metal water piping system, including the hot-water piping, and also to the sewer, gas piping, air tracts, etc. This will provide additional safety (see NEC, *Section 250.104(A)(2)*).

7-75 When reinforcing bar is used as the grounding electrode, are clamps required to be listed for connecting the grounding conductor to the rebar?

Yes. Exothermic welding is also permitted (see NEC, *Section 250.8*).

7-76 Can solder be used to attach connections to the grounding conductor?

Solder is never permitted; exothermic welding or listed pressure connectors are to be used (see NEC, *Section 250.8*).

7-77 What means must be taken to maintain continuity at metal boxes when nonmetallic systems of wiring are used so that the equipment ground wire will be continuous?

The equipment grounds must be attached by means of a grounding screw (boxes are now available with a tapped hole having 10/32 threads) or by some other listed means (see NEC, *Section 250.148*).

7-78 Are sheet-metal straps considered adequate for grounding a telephone system?

Yes, if attached to a rigid metal base that is seated on the water pipe or other ground electrode and listed for the purpose (see NEC, *Section 250.70*).

7-79 What special precautions must be taken in the use of ground clamps?

They must be made of a material suitable for use in connection with the materials that they are attaching to because electrolysis may occur if they are made of different metals. All surfaces must be clean and free of paint or corrosion (see NEC, *Section 250.70*).

7-80 May aluminum grounding conductors be run to cold water piping?

Yes, but any clamps used with aluminum grounding conductors must be approved for the purpose (see NEC, *Section 250.70*).

7-81 May conductors of different systems occupy the same enclosure?

Conductors for light and power systems of 600 volts or less may occupy the same enclosure. The individual circuits may be ac or dc, but the conductors must all be insulated for the maximum voltage of any conductor in the enclosure except emergency systems (see NEC, *Section 300.3(C)(1)*).

7-82 May conductors of systems over 600 volts occupy the same enclosure as conductors carrying less than 600 volts?

Under certain conditions (see NEC, *Section 300.3(C)(2)*).

7-83 If the secondary voltage on electric-discharge lamps is 1000 volts or less, may its wiring occupy the same fixture enclosure as the branch-circuit conductors?

Yes, if insulated for the secondary voltage involved (see NEC, *Section 300.3(C)(2)(a)*).

7-84 Is it permissible to run control, relay, or ammeter conductors that are used in connection with a motor or starter in the same enclosure as the motor-circuit conductors?

Yes, if the insulation of all the conductors is enough for the highest voltage encountered (see NEC, *Section 300.3(C)(2)(c)*).

7-85 When boxes, fittings, conduit, etc., are used in damp or corrosive places, how must they be protected?

They must be protected by a coating of approved corrosion-resistant material (see NEC, *Section 300.6*).

7-86 In damp locations, what precautions must be taken against the corrosion of boxes, fitting, conduit, etc.?

There must be an air space of at least $\frac{1}{4}$ inch between the wall or supporting material and any surface-mounted conduit fittings, etc. (see NEC, *Section 300.6(C)*).

7-87 When raceways extend from an area of one temperature into an area of a widely different temperature, what precautions must be taken?

Precautions must be taken to prevent the circulation of air from a warmer to a colder section through the raceway (see NEC, *Section 300.7(A)*).

7-88 When raceways and conductors are run through studs, joints, and/or rafters, what precautions must be taken?

They should be run near the appropriate center of the wood members or at least 1¼ inches from the nearest edge; if the members have to be notched or a 1¼-inch protection cannot be given, the conduit or conductors must be covered by a steel plate not less than 1/6 inch in thickness (see NEC, *Section 300.4*).

7-89 On 240/120-volt multiwire branch circuits feeding through an outlet box, may the neutral be broken and connected by means of the screws on the receptacle?

No. In multiwire circuits, the continuity of a grounded conductor will not be dependent upon a connection device, such as a lampholder, receptacle, etc., where the removal of such devices would interrupt the continuity (see NEC, *Section 300.13(B)*).

7-90 How much wire must be allowed at outlets and switch boxes for connections and splices?

There must be at least 6 inches of free conductor left for making the connections (see NEC, *Section 300.14*). It is a good practice, however, to leave more than 6 inches of free conductor at each box.

7-91 What precautions must be taken to prevent induced currents in metal enclosures?

When conductors carry ac, all phase wires and the neutral wire, if one is used, and all equipment grounding conductors must run in the same raceway. When single conductors must be passed through metal having magnetic properties (iron and steel), slotting the metal between the holes will help keep down the inductive effect (see NEC, *Section 300.20*).

7-92 When it is necessary to run wires through air-handling ducts or plenums, what precautions must be taken?

The conductors must be run in conduit, electrical metallic tubing, intermediate metal conduit, flexible steel conduit with lead-covered conductors, metal-clad cable, Type MI cable, intermediate metal conduit, or in a Plenum-rated cable (see NEC, *Section 300.22(B)*).

7-93 Does temporary wiring, such as for construction, Christmas lighting, carnivals, etc., come under the NEC?

Yes (see NEC, *Article 527*).

7-94 Are ground-fault circuit interrupters required on temporary wiring for construction sites?

Yes, on all 15- and 20-ampere branch circuits (see NEC, *Section 527.6(A)*).

7-95 When conductors are used underground in concrete slabs or other masonry that comes in direct contact with the earth, or where

condensation or accumulated moisture in raceways is apt to occur, what characteristic must the insulation have?

It must be suitable for use in a wet location (see NEC, *Sections 310.7* and *310.8*).

7-96   Name some insulations permitted for use in a wet location.

RHW, TW, THW, THWN, and XHHW type cable (see NEC, *Section 310.8*).

7-97   What are the cable requirements for buried conductors?

Cables of one or more conductors for direct burial in the earth must be Type USE cable, except for branch circuits and feeders, which may use Type UF cable. Type UF cable cannot be used for service wires (see NEC, *Articles 338* and *340*).

7-98   What other rules apply to buried conductors?

All conductors, including the neutral, must be buried in the same trench and be continuous (without any splice). Extra mechanical protection may be required, such as a covering board, concrete pad, or raceway (see NEC, *Section 300.5*).

7-99   What is the minimum size of conductors allowed by the NEC?

For power work, the smallest size is No. 14. However, smaller sizes are allowed for controls and in special circumstances (see NEC, *Section 310.5*).

7-100   What is the ruling on stranded conductors?

Except for bus bars, Type MI cable, and pool bonding, conductors of No. 8 and larger must be stranded (see NEC, *Section 310.3*).

7-101   What is the minimum size of conductors that may be paralleled?

Conductors in sizes 1/0 and larger may be run in multiple, though there are exceptions (see NEC, *Section 310.4*).

7-102   When conductors are run in parallel, what factors must be considered?

Each phase or neutral, if used, must be the same length and of the same conductor material; have the same circular-mil area and the same type of insulation; and be arranged to terminate at both ends so that there will be equal distribution of current between the conductors. Phases must be coordinated to eliminate any induction currents that may be caused to flow in the raceways (see NEC, *Section 310.4*).

7-103 How many conductors may be run in raceways or cables without having to apply a derating factor to the current-carrying capacity?

Three current-carrying conductors (neutrals not included); when more than three current-carrying conductors are run, a derating factor must be applied (see NEC, *Notes to Tables 310.16 through 310.19*, which is *Section 310.15(B)(2)*).

7-104 What are the derating factors for more than three current-carrying conductors in raceways or cables?

4 to 6 conductors	80% rating
7 to 9 conductors	70% rating
10 to 24 conductors	70% rating
25 to 42 conductors	60% rating
43 conductors and above	50% rating

(See NEC, *Section 310.15(B)(2)(a)*.)

7-105 In derating, how is the neutral conductor considered?

Normally, the current in the neutral conductor is only the unbalanced current. Therefore, if the system is well balanced, the neutral is not considered as a current-carrying conductor and would not enter into derating (see NEC, *Section 310.15(B)(4)*).

7-106 In determining the current in the neutral conductor of a wye system, how is the neutral classed?

In a 4-wire, 3-phase, wye-connected system, a common conductor carries approximately the same current as the other conductors and is therefore not considered a neutral in determining the derating of current capacity (see NEC, *Section 310.15 (B)(4)(c)*).

7-107 What size must the conductors be for a parallel feeder or service?

At least 1/0 and larger for aluminum, copper-clad aluminum, or copper conductors (see NEC, *Section 310.4*).

7-108 What condition permits the use of parallel conductors in sizes smaller than No. 1/0 AWG?

When run for frequencies of 360 hertz and higher (see NEC, *Section 310.4*).

7-109 What are the minimum sizes for conductors when operated at a voltage of 0 through 2000 volts?
No. 14 copper, No. 12 aluminum or copper-clad aluminum, with exceptions (see NEC, *Section 310.5*).

7-110 Where installed in raceways, stranded conductors of what size must be used?
No. 8 and larger (see NEC, *Section 310.3*).

7-111 What are the types of conductors required when installed in a wet location?
Lead-covered, types RHW, TW, THW, THWN, XHHW, or types that are listed for wet locations.

7-112 What are the temperature limitations of conductors?
See NEC, *Section 310.10*.

7-113 What are the principal determinants of conductor operating temperatures?
The maximum temperature that the conductor can withstand over a prolonged period of time without serious degradation (see NEC, *Section 310.10*, FPN).

7-114 How are grounded conductors identified?
They must be white or natural gray in color (see NEC, *Sections 310.12(A)* and *200.6*).

7-115 How are equipment grounding conductors identified?
They must be green, green with a yellow stripe, or bare (see NEC, *Sections 310.12(B)* and *250.119*).

7-116 How are ungrounded conductors identified?
They must be distinguished by colors other than white, gray, or green (see NEC, *Section 310.12(C)*).

7-117 What does the term "electrical ducts" mean?
Suitable electrical raceways, run underground and embedded in concrete or earth (see NEC, *Section 310.60(A)*).

7-118 What table covers the ampacity of type ac cable?
Table 310.16.

7-119 What is the ampacity of a No. 6 AWG copper type THW conductor?
65 amperes (see NEC, *Table 310.16*).

7-120 What are the ampacity correction factors for a Type ac cable assembly when the ambient temperature is other than 30°C (86°F)?

See NEC, *Table 310.16.*

7-121 What is the maximum operating temperature for a Type THWN conductor?

75°C (167°F) in wet locations (see NEC, *Table 310.13*).

7-122 Under what conditions can a Type XHHW conductor be used in a wet location?

When it is rated at a temperature of 75°C (167°F) (see NEC, *Section 310.13*).

7-123 What are the requirements when more than one calculated or tabulated ampacity could apply for a given circuit length?

The lowest value shall be used (see NEC, *Section 310.15 (A)(2)*).

7-124 What part of the NEC is used to determine the ampacity of underground runs of conductors?

Table 310.16.

7-125 Give the ampacities for Types ac or NM conductor assemblies based on an ambient air temperature of 30°C (86°F).

See NEC, *Table 310.16.*

7-126 What advantages would two parallel 500-MCM cables have over one 1000-MCM cable?

They would be easier to handle and to pull into raceways. According to NEC, Chapter 9, Table 8, 1000-MCM Type RH wire has a current-carrying capacity of 545 amperes, whereas 500-MCM cable has a current-carrying capacity of 380 amperes. Therefore, two 500-MCM cables in parallel would have the same circular-mil area as one 1000-MCM cable and a current-carrying capacity of 760 amperes compared to the 545-ampere current-carrying capacity of one 1000-MCM cable. This is approximately 40 percent more current-carrying capacity for the two 500-MCM cables. In addition, there is usually a cost advantage by using two 500-MCM conductors, rather than one 1000-MCM conductor.

7-127 May THW insulated conductors be run through a continuous row of fluorescent fixtures?

Yes (see NEC, *Table 310.13* and *Section 410.31*).

Chapter 8

Wiring Methods and Materials

8-1   Can cable trays be used to support welding cables?
Yes. They must be installed in accordance with NEC, *Section 630.42* (see NEC, *Section 392.3(B)(1)(b)*).

8-2   Are cable trays intended to be used with ordinary branch circuit conductors (as described in *Article 392* of the NEC)?
No. They are not intended for this purpose (see NEC, *Section 392.3(B)(1)*).

8-3   Do cable trays have to be used as a complete system, or may they be used as only partial cable supports?
They must be used as a complete system, including boxes and fittings (see NEC, *Section 392.6(A)*).

8-4   Is grounding required on metallic cable trays?
They must be grounded but must not be used as the neutral conductor. If used as the equipment-grounding conductor, they must meet the requirements of *Section 392.7*.

8-5   Is open wiring on insulators, commonly known as knob-and-tube wiring, recognized by the National Electrical Code?
Yes, although it is almost never approved by inspectors for new installations. It is in the NEC primarily because a great deal of it has been previously installed and must be properly maintained (see NEC, *Section 394.10*).

8-6   What is Type MI cable?
Type MI cable is a cable in which one or more electrical conductors are insulated with a highly compressed refractory mineral insulation (magnesium oxide) and enclosed in a liquid-tight and gas-tight metallic sheathing (copper) (see NEC, *Section 332.2*).

8-7   Where can Type MI cable be used?
This is one wiring material that can be used for practically every conceivable type of service or circuit. When it is exposed to cinder fill or other destructive corrosive conditions, it must be protected by materials suitable for these conditions (see NEC, *Section 332.10*).

8-8   What precautions must be taken with Type MI cable to prevent the entrance of moisture?

Where Type MI cable terminates, an approved seal must be provided immediately after stripping to prevent the entrance of moisture into the mineral insulation; the conductors must be insulated with an approved insulation where they extend beyond the sheath (see NEC, *Section 332.40(B)*).

8-9   Does the outer sheath of Type MI cable meet the requirements for equipment grounding?

Yes. It provides an excellent path for equipment grounding purposes but not for use as a grounded or neutral conductor (see NEC, *Sections 332.108*).

8-10   What is armored cable commonly called?

Type AC cable, or in some locations "BX," a name taken from the Bronx Cable Company (see NEC, *Article 320*).

8-11   What is metal-clad cable commonly called?

Type MC cable (see NEC, *Article 330*).

8-12   What is Type MC cable?

Type MC cables are power and control cables in the size range from No. 14 and larger for copper and No. 12 and larger for aluminum and copper-clad aluminum.

The metal enclosures are either a covering of interlocking metal tape or a smooth, impervious, close-fitting, or corrugated tube. Supplemental protection of an outer covering of corrosion-resistant material is required where such protection is needed (see NEC, *Sections 330.2 and 330.116*).

8-13   What is Type AC cable?

Type AC cables are branch-circuit and feeder cables with an armor of flexible metal tape. Cable of the AC type must have an internal bonding strip of copper or aluminum in tight physical contact with the armor for its entire length (see NEC, *Section 320.2*).

8-14   At what intervals must Type MC cable be supported?

At intervals not exceeding 6 feet (see NEC, *Section 330.30*).

8-15   At what intervals must Type AC cable be supported?

At intervals of not more than 4¼ feet and within 12 inches of boxes and fittings, except where cable is fished and except lengths of not over 2 feet at terminals where flexibility is necessary, and in lengths not over 6 feet for fixture whips (see NEC, *Section 320.30*).

8-16 How may bends be made in Type AC cable?

Bends must be made so that the cables are not damaged. The radius of the curve of the inner edge of any bend must not be less than 5 times the diameter of Type AC cable (see NEC, *Section 320.24*).

8-17 What extra precaution is necessary with Type AC cable when attaching fittings?

An approved insulating bushing or equivalent approved protection must be provided between the conductors and the armor (see NEC, *Section 320.40*).

8-18 What is nonmetallic-sheathed cable?

Nonmetallic-sheathed cable is an assembly of two or more insulated conductors having an outer sheath of moisture-resistant, flame-retardant, nonmetallic material (see NEC, *Section 334.2*).

8-19 What are the available sizes of nonmetallic-sheathed cable?

Sizes No. 14 to No. 2 (AWG) inclusive (see NEC, *Section 334.104*).

8-20 What is the difference between Type NM and Type NMC cables?

In addition to having flame-retardant and moisture-resistant covering, Type NMC cable must also be fungus-resistant and corrosion-resistant (see NEC, *Section 334.116(B)*).

8-21 Where may Type NM cable be used?

It may be used for exposed and concealed work in normally dry locations. It may also be installed or fished in the air voids of masonry block or tile walls, where such walls are not exposed to excessive moisture or dampness (see NEC, *Section 334.10(A)*).

8-22 In dwelling occupancies, what, if any, restrictions have been added?

Types NM and NMC cables are permitted in one- and two-family dwellings or multifamily dwellings and other structures not exceeding three floors above grade (see NEC, *Sections 334.10(2)* and *334.10(3)*).

8-23 Where may Type NMC cable be installed?

Where exposed to corrosive fumes or vapors, for exposed and concealed work, in dry and moist or damp places, and inside or outside of masonry block or tile walls (see NEC, *Section 334.10*).

8-24 Where is it not permissible to use Type NM or Type NMC cable?

As service-entrance cable, in commercial garages, in theaters, and similar locations, except as provided in the NEC, *Article 518*, "Places of Assembly" (see NEC, *Section 334.12*).

8-25 What supports are necessary for Types NM and NMC cable?

The installation must not be subject to damage, and the cable must be secured in place at intervals not exceeding 4½ feet and supported within 12 inches from boxes and fittings (see NEC, *Section 334.30*).

8-26 What is service-entrance cable?

Service-entrance cable is a conductor assembly provided with a suitable covering, primarily used for services (see NEC, *Section 338.2*).

8-27 What are the two types of service-entrance cable?

Type SE and Type USE (see NEC, *Section 338.2*).

8-28 What type of outer sheath does a Type SE cable have?

It has a flame-retardant, moisture-resistant covering (see NEC, *Section 338.1*).

8-29 What type of outer sheath does Type USE cable have?

For underground use, it has a moisture-resistant covering but is not required to have a flame-retardant covering (see NEC, *Section 338.1*).

8-30 Can service-entrance cable be used as branch-circuit conductors or feeder conductors?

It may be used for interior wiring systems only when all circuit conductors of the cable are of the thermoplastic or rubber type. However, service-entrance cable without individual insulation on the grounded (neutral) conductor can be used to supply ranges or clothes dryers when the cable does not have an outer metallic covering and does not exceed 150 volts to ground. It may also be used as a feeder when it supplies other buildings on the same premises. SE cable with a bare neutral must not originate in a feeder panel, as covered in *Section 250.140* (see NEC, *Section 338.10(B)*).

8-31 Describe underground feeder and branch-circuit cable.
Underground feeder and branch-circuit cable must be an approved Type UF cable in sizes No. 14 copper or No. 12 aluminum or copper-clad aluminum to No. 4/0 AWG, inclusive. The ampacity of Type UF cable shall be that of 60°C (140°F) conductors in accordance with *Section 310.13*. In addition, the cable can include a bare conductor for equipment grounding purposes only. The overall covering must also be suitable for direct burial in the earth (see NEC, *Sections 340.2*).

8-32 What is required as overcurrent protection for Type UF cable?
Underground feeder and branch-circuit cable may be used underground, including direct burial, provided they have overcurrent protection that does not exceed the capacity of the conductors (see NEC, *Section 340.80*).

8-33 May conductors of the same circuit be run in separate trenches?
All conductors of the same circuit must be run in the same trench or raceway (see NEC, *Sections 340.10(2)* and *300.3(B)*).

8-34 What depth must UF cable be buried?
In general, UF cable must be buried at a depth of 24 inches or more. However, there are circumstances under which it can be buried at lesser depths (see NEC, *Section 340.10* and *Table 300.5*).

8-35 May UF cable be used for interior wiring?
Yes, but it must be installed in accordance with *Article 334* and must not be exposed to direct sunlight unless it is specifically identified for this purpose (see NEC, *Section 340.10(4)*).

8-36 What is the normal taper on a standard conduit thread-cutting die?
¾ inch per foot (see NEC, *Section 344.28*).

8-37 Where can rigid metal conduit be used?
Practically everywhere; however, rigid metal conduit protected solely by enamel cannot be used outdoors. Also, where practicable, dissimilar metals should not come in contact anywhere in the system to avoid the possibility of galvanic action, which could prove destructive to the system (see NEC, *Section 344.10*).

8-38 When subject to cinder fill, what precautions must be taken with rigid metal conduit?
It must be protected by a minimum of 2 inches of noncinder concrete (see NEC, *Section 344.10(C)*).

8-39 What is the minimum size of rigid metal conduit?
½-inch trade size, with a few exceptions (see NEC, *Section 344.20(A)*).

8-40 What are the requirements for couplings on rigid metal conduit?
Threaded or threadless couplings and connectors must be made tight. Where installed in wet locations or buried in concrete masonry or fill, the type used must prevent water from entering the conduit (see NEC, *Section 344.42*).

8-41 Are running threads permitted on rigid metal conduit?
They cannot be used on conduit for connections at couplings (see NEC, *Section 344.42(B)*).

8-42 What is the maximum number of bends permitted in rigid metal conduit between outlets?
Not more than 4 quarter bends, or a total of 360°; this includes offsets, etc., at outlets and fittings (see NEC, *Section 344.26*).

8-43 What is the minimum radius of conduit bends?
See NEC, *Table 344.24*.

8-44 After a cut is made in rigid metal conduit, what precaution must be taken?
All cut ends of conduits must be reamed to remove any rough ends that might damage the wire when it is pulled in (see NEC, *Section 344.28*).

8-45 What number of conductors are permitted in conduit?
See *Annex* C of the NEC.

8-46 What are the temperature limitations of conductors?
See NEC, *Section 310.10*.

8-47 What are the principal determinants of conductor operating temperatures?

The maximum temperature that the conductor can withstand over a prolonged period of time without serious degradation (see NEC, *Section 310.10*, FPN).

8-48 How are grounded conductors identified?
They must be white or natural gray in color (see NEC, *Section 310.12(A)*).

8-49 How are equipment grounding conductors identified?
They must be green, green with a yellow stripe, or bare (see NEC, *Sections 310.12(B)* and *250.119*).

8-50 How are ungrounded conductors identified?
They must be distinguished by colors other than white, gray, or green (see NEC, *Sections 310.12(C)* and *200.7*).

8-51 What does the term "electrical ducts" mean?
Suitable electrical raceways, run underground and embedded in concrete or earth (see NEC, *Section 310.60*).

8-52 What table covers the ampacity of Type AC cable?
Table 310.16.

8-53 What is the ampacity of a No. 6 AWG copper Type THW conductor?
65 amperes (see NEC, *Table 310.16*).

8-54 What are the ampacity correction factors for a Type AC cable assembly when the ambient temperature is other than 30°C (86°F)?
See NEC, *Section 310.16*.

8-55 What is the maximum operating temperature for a Type THWN conductor?
75°C (167°F) (see NEC, *Section 310.13*).

8-56 Under what conditions can a Type XHHW conductor be used in a wet location?
When it is rated at a temperature of 75°C (167°F) (see NEC, *Section 310.13*).

8-57 What are the requirements when more than one calculated or tabulated ampacity could apply for a given circuit length?
The lowest value shall be used (see NEC, *Section 310.15 (A)(2)*).

8-58 What table is used for directly buried conductors?
Table 310.16.

8-59 What is intermediate metal conduit?
It is a metal conduit, similar to rigid metal conduit, but slightly lighter weight (see NEC, *Article 342*).

8-60 What is the minimum size for intermediate metal conduit?
½-inch electrical trade size (see NEC, *Section 342.20(A)*).

8-61 What is the maximum size for intermediate metal conduit?
4-inch electrical trade size (see NEC, *Section 342.20(B)*).

8-62 What are the support requirements for runs of intermediate metal conduit?
See NEC, *Section 342.30*.

8-63 What are the marking requirements for intermediate metal conduit?
See NEC, *Section 342.120*.

8-64 What precautions should be taken when installing rigid non-metallic conduit in extreme cold?
They should not be used in areas where they will be subjected to damage, since they can become brittle at low temperatures (see NEC, *Section 352.10* FPN).

8-65 Should rigid nonmetallic conduit be of a suitable type?
Yes. It must be listed (see NEC, *Section 352.6*).

8-66 May plastic water pipe be used as conduit?
No. It may not be used for this purpose (see NEC, *Section 110.3(B)*).

8-67 What uses are permitted for rigid nonmetallic conduit?
For 600 volts or less, except for direct burial where it is not less than 18 inches below grade and where the voltage exceeds 600 volts, in which case it must be encased in 2 inches of concrete; in concrete walls, floors, and ceilings; in locations subject to severe corrosive influences; in cinder fill; and in wet locations (see NEC, *Article 352.10*).

8-68 Where is the use of nonmetallic rigid conduit prohibited?
In hazardous locations, for the support of fixtures or other equipment, where subject to physical damage, or where high temperatures are present (see NEC, *Section 352.12*).

8-69 Where must expansion joints be used on rigid nonmetallic conduit?

Where required to compensate for thermal expansion and contraction when the expansion, as determined by *Table 352.44 (A) or (B)*, will be ¼ inch or greater between terminations (see NEC, *Section 352.44*).

8-70 What is the minimum size of rigid nonmetallic conduit?

No conduit smaller than ½-inch electrical trade size may be used (see NEC, *Section 352.20(A)*).

8-71 Are bushings required when rigid nonmetallic conduit is used?

Yes (see NEC, *Section 352.46*).

8-72 What is electrical metallic tubing (EMT)?

A thin-walled metal raceway (see NEC, *Article 358*).

8-73 May EMT be threaded?

No. EMT can be coupled only by suitable fittings. Where integral couplings are utilized, such couplings shall be permitted to be factory threaded (see NEC, *Section 358.28(B)*).

8-74 In what minimum and maximum sizes is EMT available?

The minimum size is ½ inch, with exceptions, and the maximum size is 4 inches, electrical trade size (see NEC, *Section 358.20*).

8-75 May more or fewer conductors be contained in EMT than in rigid metal conduit?

The same number of wires applies to both raceways (see NEC, *Sections 358.22 and 344.22*).

8-76 How do the number of bends and the reaming requirements for EMT compare to rigid metal conduit?

They are the same for both (see NEC, *Sections 358.24, 358.28, 344.24, and 344.28*).

8-77 When EMT is exposed to moisture or used outdoors, what types of fittings must be used?

They must be rain-tight fittings (see NEC, *Section 358.42*).

8-78 When EMT is buried in concrete, what type fittings must be used?

They must be concrete-tight fittings (see NEC, *Section 358.42*).

8-79 What are some common names for flexible metal conduit?

Flex, Greenfield (see NEC, *Index*).

8-80 What is the minimum size for flexible metal conduit?
The normal minimum size is ½ inch, although there are some exceptions to this (see NEC, *Section 348.20(A)*).

8-81 What is the number of wires permitted in flexible metal conduit?
The requirement for this is the same as for rigid metal conduit; however, when ⅜-inch conduit is permitted, a special conductor table is required (see NEC, *Section 348.22*).

8-82 What are the sizes of liquid-tight flexible metal conduit?
The sizes of liquid-tight flexible metal conduit must be electrical trade sizes ½ to 4 inches, inclusive. Exception: ⅜-inch size may be used as permitted in the NEC, *Section 350.20*.

8-83 How often should liquid-tight flexible metal conduit be supported?
Where liquid-tight flexible metal conduit is installed as a fixed raceway, it shall be secured by approved means at intervals not exceeding 4½ feet and within 12 inches on each side of every outlet box or fitting except where conduit is fished, where flexibility is required, and for fixture whips (see NEC, *Section 350.30*).

8-84 May regular flexible metal conduit fittings be used with liquid-tight flexible metal conduit?
No. They must be fittings identified for use with liquid-tight flexible metal conduit (see NEC, *Section 350.6*).

8-85 What is liquid-tight flexible nonmetallic conduit?
See NEC, *Section 356.2*.

8-86 Where can liquid-tight flexible nonmetallic conduit be used in lengths exceeding 6 feet?
Where it is a pre-wired assembly, supported in accordance with *Section 356.30*. Also see NEC, *Section 356.100*.

8-87 Under what conditions can liquid-tight flexible nonmetallic conduit be used in lengths exceeding 6 feet?
See NEC, *Section 351-23(B)(3)*, Exception.

8-88 May liquid-tight flexible metal conduit be used as an equipment ground?
Liquid-tight flexible metal conduit may be used for grounding in the 1¼-inch and smaller trade sizes if the length is 6 feet or less

and it is terminated in fittings listed for grounding (see NEC, *Section 250.118(7)*).

8-89 What are metal wireways?

Metal wireways are sheet-metal troughs with hinged or removable covers for housing and protecting electric wires and cable and in which conductors are laid in place after the wireway has been installed as a complete system (see NEC, *Section 376.2*).

8-90 What use of metal wireways is permitted?

Metal wireways may be installed only for exposed work; where wireways are intended for outdoor use, they must be of rain-tight construction (see NEC, *Section 376.10*).

8-91 What are the uses for which metal wireways are prohibited?

Where subject to severe physical damage or corrosive vapors and in hazardous (classified) locations (see NEC, *Section 376.12*).

8-92 What is the largest size wire permitted in wireways?

No conductor larger than that for which the wireway is designed may be installed in any wireway (see NEC, *Sections 376.21 and 378.21*).

8-93 What is the maximum number of conductors permitted in metal wireways, and what is the maximum percentage of fill?

The maximum fill cannot exceed 20 percent of the interior cross-sectional area. A maximum of 30 current-carrying conductors at any cross section (unless derated); signal circuits or starter-control wires are not included (see NEC, *Section 376.22*).

8-94 Are splices permitted in wireways?

Yes, provided that the conductors with splices do not take up more than 75 percent of the area of the wireway at any point (see NEC, *Sections 376.56 and 378.56*).

8-95 May the wireways be open at the ends?

No. The dead ends must be closed (see NEC, *Sections 376.58 and 378.58*).

8-96 Is the equipment-grounding conductor used in figuring the number of conductors allowed in a box?

Yes, but only one conductor is to be counted (see NEC, *Section 314.16(B)(5)*).

8-97 What is the purpose of auxiliary gutters?

They are used to supplement wiring spaces at meter centers, distribution centers, switchboards, and similar points of wiring systems; they may enclose conductors or bus bars but cannot be used to enclose switches, overcurrent devices, or other appliances or apparatuses (see NEC, *Section 366.2*).

8-98 What is the maximum number of conductors permitted in auxiliary gutters, and what is the maximum percentage of fill allowed?

The maximum fill cannot exceed 20 percent of the interior cross-sectional area. A maximum of 30 current-carrying conductors at any cross section (unless derated); signal circuits or starter-control wires are not included (see NEC, *Section 366.6*).

8-99 What is the current-carrying capacity of bus bars in auxiliary gutters?

The current carried continuously in bare copper bars in auxiliary gutters shall not exceed 1000 amperes per square inch of cross section of the conductor. For aluminum bars the current carried continuously shall not exceed 700 amperes per square inch of cross section of the conductor (see NEC, *Section 366.7*).

8-100 May taps and splices be made in auxiliary gutters?

Yes; however, splices may not occupy more than 75 percent of the cross-sectional area at any point. All taps must be identified as to the circuit or equipment that they supply (see NEC, *Section 366.9(A)*).

8-101 May round boxes be used as outlet, switch, or junction boxes?

No. They cannot be used where conduits or connectors with locknuts or bushings are used (see NEC, *Section 314.2*).

8-102 Is the number of conductors permitted in outlet, switch, and junction boxes limited to any certain number?

Yes; see NEC, *Tables 314.6(A)* and *31470.6(B)*.

Any box less than 1½ inches deep is considered to be a shallow box.

These tables are quite plain; however, they are for conduit or where connectors are used. If cable clamps are used, one conductor must be deducted from the table to allow for the clamp space. Where one or more fixture studs, cable clamps, or hickeys are contained in the box, the number of conductors will be one less than shown in the tables, with a further deduction of one conductor for one or several flush devices mounted on the same strap. A double receptacle is one device; a single switch is one device; three despard switches on one strap is one device; and three despard receptacles on one strap is one device.

There are times when you will have combinations of wire sizes or other conditions, and you won't find the answer in *Table 370.6(A)*. *Table 370.6(B)* gives the volume of free space (in cubic inches) within a box that is required for each wire size specified. Determine the number of cubic inches in the box; from this you can find the number of various combinations of wire.

8-103 May nonmetallic boxes be used?
Yes, with a number of restrictions (see NEC, *Section 314.3*).

8-104 How are metal boxes grounded?
In accordance with the requirements of *Article 250* (see NEC, *Section 314.4*).

8-105 How would you determine the proper size for a square box containing No. 18 or No. 16 conductors?
By referring to *Table 314.6(A)*.

8-106 What is the cubic inch capacity required for a No. 18 conductor?
1.5 cubic inch (see NEC, *Table 314.6(B)*).

8-107 What are the support requirements for boxes and fittings?
See NEC, *Section 314.23*.

8-108 When are exposed surface extensions permitted?
See NEC, *Section 314.22*.

8-109 What are the requirements for pendant boxes?
Pendant boxes must be supported by either multiconductor cords or cables in an approved manner, or by conduit. See NEC, *Section 314.23(H)*.

8-110 Can a box be used to support a ceiling fan?
Yes (see NEC, *Sections 314.27(D)* and *422.18*).

8-111 What protection must be given to conductors entering boxes or fittings?
They must be protected from abrasion, and the openings through which conductors enter must be adequately closed (see NEC, *Section 314.17*).

8-112 May unused openings in boxes and fittings be left open?

No. They must be adequately closed with protection equivalent to that of the wall of the box or fitting (see NEC, *Section 314.17(A)*).

8-113 How far may boxes be set back in walls?

In walls and ceilings of concrete or other noncombustible materials, the boxes may be set back not more than ¼ inch. In walls or ceilings of wood or other combustible materials, the boxes must be flush with the finished surface. If the plaster is broken or incomplete, it must be repaired, so that there will be no gaps or openings greater than ⅛ inch around the box (see NEC, *Sections 314.20* and *314.21*).

8-114 How can you determine dimensions for pull or junction boxes?

For raceways of ¾-inch trade size and larger containing conductors of No. 4 wire or larger, the minimum dimensions of a pull or junction box installed in the raceway for straight pulls must not be less than eight times the trade diameter of the largest raceway. For angle or U pulls, the distance between each raceway entry inside the box and opposite wall of the box must not be less than six times the trade size of the raceway. The distance must be increased for additional entries by the amount of the sum of the diameters of all other raceway entries on the same wall of the box. The distance between raceway entries enclosing the same conductor cannot be less than six times the electrical trade diameter of the larger raceway (see NEC, *Section 314.28*).

8-115 Must pull, outlet, and junction boxes be accessible?

Yes, they must be accessible without removing any part of the building, sidewalk, or paving (see NEC, *Section 314.29*).

8-116 When installing cabinets and cutout boxes in damp or wet places, what precautions must be taken?

They must be installed so that any accumulated moisture will drain out. There must also be a ¼-inch air space between the enclosure and the surface on which they are mounted (see NEC, *Section 312.2(A)*).

8-117 When mounting cabinets and cutout boxes, how much setback is allowed?

If mounted in concrete or other noncombustible material, they must not be set back more than ¼ inch. If mounted in wood or other combustible material, they must be mounted flush with the finished surface (see NEC, *Section 312.3*).

8-118 Is there any limitation of the deflection of conductors at terminal connections?
Yes. Use Table 312.6(A) of the NEC. If conductors are deflected or bent too sharply, damage or breakage of conductors may result (see NEC, *Section 312.6*).

8-119 Can unused openings in cabinets or cutout boxes be left open?
No. They must be effectively closed (see NEC, *Section 312.5(A)*).

8-120 What type of bushings must be used on conduits with No. 4 wire or larger?
The bushings must be of the insulating type, or other approved methods must be used when insulating the bushings. Where bonding bushings are also required, combination bonding and insulating bushings are available (see NEC, *Sections 312.6(C)* and *300.4(F)*).

8-121 Are meter sockets covered by the NEC?
Yes (see NEC, *Section 312.1*).

8-122 What must be done where conductors enter a cabinet or cutout box and are within a cable assembly?
Each cable must be secured to the cabinet or cutout box (see NEC, *Section 312.5(C)*).

8-123 What does the term "offset" mean, as applied to conductors within enclosures?
See NEC, *Section 312.6(2)* FPN.

8-124 May enclosures for switches or overcurrent devices be used on raceways or junction boxes?
Enclosures for switches or overcurrent devices must not be used as junction boxes, auxiliary gutters, or raceways for conductors feeding through or tapping off to other switches or overcurrent devices, unless adequate space is provided so that the conductors don't fill the wiring space at any cross section to more than 40 percent of the cross-sectional area of the space, and so that the conductors, splices, and taps don't fill the wiring

space at any cross section to more than 75 percent of the cross-sectional area of the space (see NEC, *Section 312.8*).

8-125 May switches be connected in the grounded conductor?
No, unless the ungrounded conductor is opened at the same time as the grounded conductor (see NEC, *Section 404.2(B)*).

8-126 Are signal circuits or control conductors for a motor and starter in auxiliary gutters considered current-carrying conductors?
No (see NEC, *Section 366.6(A)*).

8-127 Can the grounded conductor be switched when using three- or four-way switches?
No. The wiring must be arranged so that all the switching is done in the ungrounded conductor. When the wiring between switches and outlets is run in metal raceways, both polarities must be in the same enclosure (see NEC, *Section 404.2(A)*).

8-128 In what position must knife switches be mounted?
Single-throw knife switches must be mounted so that gravity won't tend to close them. Double-throw knife switches may be mounted either vertically or horizontally; however, if mounted vertically, a locking device must be provided to ensure that the blades remain in the open position when so set (see NEC, *Section 404.6*).

8-129 Do switches and circuit breakers have to be accessible and grouped?
Yes. Switches and circuit breakers, as far as is practical, must be readily accessible (see NEC, *Section 404.8(A)*).

8-130 How must the blades of knife switches be connected?
Unless they are the double-throw-type switches, they must be connected so that when open, the blades will be "dead" (see NEC, *Section 404.6(C)*).

8-131 May knife switches rated at 250 volts at more than 1200 amperes or 600 volts at more than 600 amperes be used as disconnecting switches?
No. They may be used for isolation switches only. For interrupting currents such as these or larger, a circuit breaker or a switch of special design listed for such purpose must be used (see NEC, *Section 404.13(A)*).

8-132 May fuses be used in parallel on switches?

No. A fused switch shall not have fuses in parallel unless fuses are factory assembled in parallel and approved as a unit (see NEC, *Section 240.8*).

8-133 What is the maximum number of overcurrent devices permitted on one panelboard?

Not more than 42 overcurrent devices of a lighting and appliance branch-circuit panelboard can be installed in any one cabinet or cutout box. A two-pole circuit breaker is considered as two overcurrent devices, and a three-pole circuit breaker is considered as three overcurrent devices in the interpretation of the question (see NEC, *Section 408.15*).

8-134 How would you define a lighting and appliance branch-circuit panelboard?

A branch-circuit panelboard that has more than 10 percent of its overcurrent devices rated 30 amps or less (see NEC, *Section 408.14(A)*).

8-135 What methods are specified in the Code for protecting conductors in trenches from damage due to ground movement?

Coiling the conductor or cable (see NEC, *Section 300.5(J)*).

8-136 Must all underground metal conduit and fittings be grounded?

Generally yes, but an exception is made for a metal conduit elbow in a rigid nonmetallic conduit run, if it is no less than 18 inches below grade, or if it is beneath a 4-inch-thick concrete slab (see NEC, *Section 250.86*).

Chapter 9

Batteries and Rectification

9-1   What is a cell (as referred to in connection with batteries)?
A cell is a single unit capable of producing a dc voltage by converting chemical energy into electrical energy.

9-2   What is a primary cell?
A primary cell is a cell that produces electric current from an electrochemical reaction but is not capable of being recharged.

9-3   What is a secondary cell?
A cell that is capable of being recharged by passing an electric current through it in the opposite direction from the discharging current.

9-4   What is a battery?
Two or more dry cells or storage cells connected together to serve as a single dc voltage source.

9-5   What are the three most common types of cells?
The lead-acid cell, the alkaline cell, and the ordinary dry cell.

9-6   What type of cell is used as an automobile battery?
The lead-acid cell is the one in common use, although the highly expensive and highly dependable nickel-cadmium battery is used to some extent.

9-7   What is the rated voltage of a standard lead-acid cell?
2 volts.

9-8   How are automobile batteries rated?
In ampere-hours.

9-9   Explain the meaning of ampere-hours.
At full charge, the battery is capable of delivering x number of amperes for y number of hours; e.g., a 100-ampere-hour battery would be capable of delivering 10 amperes for 10 hours, 1 ampere for 100 hours, etc.

9-10 What is the specific gravity of the acid in a fully charged (lead-acid) car battery?

Approximately 1.280 to 1.300.

9-11 What is the specific gravity of a discharged car battery?

Approximately 1.200 to 1.215.

9-12 Is it necessary to replace the acid in a car battery?

Under normal conditions, no.

9-13 When adding water to a car battery, what precautions should be observed?

Use distilled water only, and fill the cells only to their prescribed levels.

9-14 Will a discharged car battery freeze easier than a fully charged battery?

Yes.

9-15 Why is it necessary to occasionally recharge a lead-acid battery even though it is not being used?

A lead-acid battery not in use will gradually lose its charge, and if it is left in an uncharged condition, the material on the plates will flake off and short-circuit the plates, thereby causing a shorted cell or cells.

9-16 How can a lead-acid battery be recharged?

By connecting it to a dc source at slightly higher than battery voltage and passing a high current through it in the opposite direction from the discharging current.

9-17 How can you obtain a dc source of power from an ac power source?

By means of ac power rectification.

9-18 Name some rectifiers in common use.

Vacuum tube rectifiers and semiconductor-diode rectifiers, such as silicon and germanium.

9-19 How does a vacuum-tube rectifier operate?

A vacuum tube rectifier converts an alternating current into an unidirectional (direct) current.

9-20 How does a contact rectifier (barrier-layer rectifier) work?
It allows passage of current through the contact surface of two materials much more easily in one direction than in the other direction. The contact, or boundary, surface between the two materials is called the barrier, or blocking, layer.

9-21 Name some barrier-layer rectifiers.
Copper-oxide cells, selenium rectifier cells, magnesium-copper-sulfide cells, and semiconductor rectifiers, such as germanium and silicon.

9-22 Name two common rectifier circuits.
Half-wave rectifier and full-wave rectifier.

9-23 What is a half-wave rectifier?
A rectifier that passes only one-half of an ac wave.

9-24 What is a full-wave rectifier?
A rectifier that passes both halves of an ac wave.

9-25 Draw two cycles of an alternating current. Show the output waveform from a half-wave rectifier and the output waveform from a full-wave rectifier.
See Figure 9-1.

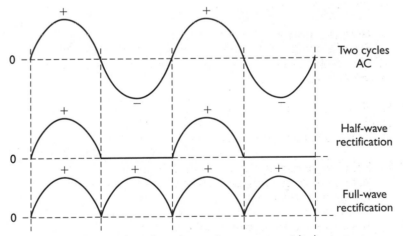

Figure 9-1 Two cycles of an alternating current, with the output waveforms from a half-wave rectifier and a full-wave rectifier.

9-26 Can the output wave from half- and full-wave rectification be smoothed out?
Yes, by the proper use of capacitors in the rectifier circuit.

9-27 Name two mechanical methods of producing direct current from alternating current.
A motor-generator set and a rotary converter.

9-28 What is the voltage of the ordinary dry cell?
1½ volts.

9-29 When more voltage is required than one cell produces, how should the cells be connected?
In series.

9-30 When more current is required than one cell can supply, how should the cells be connected?
In parallel.

9-31 How many cells are there in a 12-volt car battery, and how are they connected?
Six cells are connected in series. Each cell supplies 2 volts, and when six cells are connected in series, they supply the required 12 volts.

Chapter 10

Voltage Generation

10-1   What is a thermocouple?
A thermocouple consists of two dissimilar metals connected together at one point. When heat is applied to this point, a voltage will be generated.

10-2   Illustrate a thermocouple.
See Figure 10-1.

10-3   What are some uses of a thermocouple?
As an electric thermometer—the voltmeter connected across the circuit is calibrated in degrees of temperature. As a source of electricity on gas furnaces—for operating the valves on the furnace without an external source of power.

10-4   Our normal sources of power are generators or alternators. By what principle is power generated in these devices?
It is generated by conductors cutting magnetic lines of force, or by the lines of force cutting the conductors.

10-5   When conductors cut lines of force, what factors determine the voltage generated?
The strength of the magnetic field, the number of conductors in series, and the speed at which the field is cut.

10-6   Basically, what must a generator consist of?
A magnetic field (usually an electromagnet energized by direct current, an armature, or a rotor) coils on an iron frame, and some device for taking the current from the rotor or armature—a commutator or slip rings.

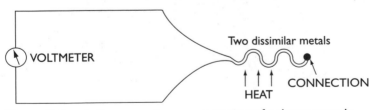

Figure 10-1   A diagrammatic representation of a thermocouple.

10-7 In a standard dc generator, what type of voltage is generated?
The voltage generated is direct current.

10-8 What is the main difference between an alternator and a dc generator?
The method by which the generated power is taken from the generator. The alternator uses slip rings, and the dc generator has a commutator, which takes the current from the coils in such a manner that it always flows in the same direction.

10-9 Draw a simple alternator.
See Figure 10-2.

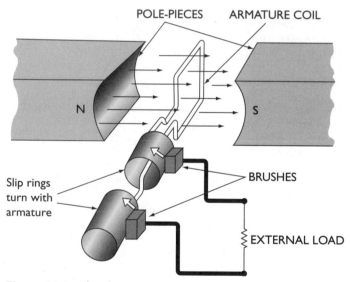

Figure 10-2 An alternator.

10-10 What does the waveform taken from an alternator look like?
See Figure 10-3.

10-11 Draw a simple dc generator with only two commutator segments.
See Figure 10-4.

10-12 What does the waveform from a commutator with only two segments look like?
See Figure 10-5.

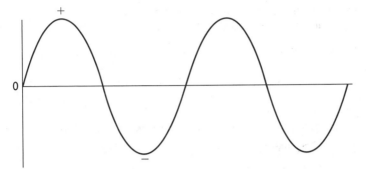

Figure 10-3   The waveform from the output of an alternator.

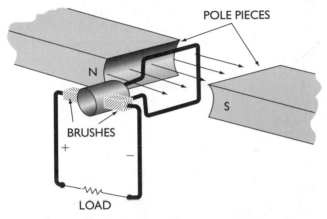

Figure 10-4   dc generator with two commutator segments.

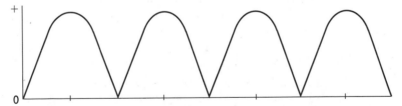

Figure 10-5   The waveform from the output of a dc generator having two commutator segments.

10-13 When more segments are added to the commutator on a dc generator, what happens to the waveform? Explain and illustrate.

The waveform tends to straighten out into a more nearly pure dc waveform (Figure 10-6).

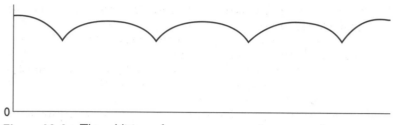

Figure 10-6 The addition of commutator segments to a dc generator has the effect of straightening out the output waveform.

10-14 How is the voltage output on a dc generator regulated?
It may be regulated by changing the strength of the field excitation by means of a rheostat, or by controlling the generator speed.

10-15 How is the voltage output of an alternator controlled?
Since the frequency of the output must be maintained constant, the field excitation must be varied with a rheostat.

10-16 What is the formula for figuring the output frequency of an alternator?

$$\text{Frequency (in hertz)} = \frac{\text{pairs of poles} \times \text{rpm}}{60}$$

10-17 What is the frequency of a 4-pole alternator, driven at 1800 rpm?
A 4-pole alternator has 2 pairs of poles. Therefore, the frequency is:

$$\frac{2 \times 1800}{60} = \frac{3600}{60} = 60 \text{ hertz}$$

10-18 How is the field on a dc generator usually excited?
It is usually self-excited. As the generator starts generating, due to residual magnetism, part of the output is fed into the field.

10-19 How is the field on an alternator usually excited?
A dc generator is usually attached to the end of the alternator shaft. It is from this generator that the excitation current for the alternator field is received.

10-20 On an alternator, does the dc part (the poles) or the ac part (the conductors) rotate?

It is immaterial which part rotates. However, the dc field is usually made the rotating part, and the stator is usually the ac part of the device. This is because the dc field excitation can be of relatively low voltage, and it is easier to insulate a rotating part for low voltage than for high voltage. The ac output is usually a much higher voltage, and it is much more practical to insulate the stator for the high voltage. Also, with this arrangement there are no brushes required on the output side.

10-21 With a rotating dc field on an alternator, what is used for the dc input?

Slip rings are used.

10-22 On a dc generator, which parts, if any, must be laminated? Why?

The armature must be laminated to lower the hysteresis and eddy-current losses. It is not necessary to laminate the field because it is strictly direct current, and there would be no hysteresis or eddy-current losses from a dc field.

10-23 On an alternator, what parts must be laminated? Why?

The stator must be laminated because of hysteresis and eddy-current losses. The rotor need not be laminated because the magnetic flux is steady.

10-24 Why is it preferable to generate alternating current rather than direct current?

Alternating current can be changed in voltage by means of transformers. This is necessary because to transmit power over any distance it must be at high voltage. Too much power is lost when transmitting at low voltage. Direct current cannot be changed in voltage without first changing it to alternating current and then raising the voltage; the operation must then be reversed at the receiving end.

10-25 Is it possible to generate power with a common three-phase, squirrel cage motor?

Yes. This is known as an induction generator.

10-26 How can a common three-phase, squirrel cage motor be made to generate?

It must be connected in a three-phase circuit and be driven by a prime mover at a faster-than-synchronous speed. It won't generate unless it is connected in a circuit that is already supplying current, or is given a short burst of current to the windings.

10-27 With a squirrel cage motor connected to line and run above synchronous speed (as described in 10-26), will the motor develop wattless power or true power?

It is capable of supplying only true power. The alternator must supply the wattless power component.

Chapter 11

Equipment for General Use

11-1 What may flexible cord be used for?
Flexible cord may be used only for pendants, wiring of fixtures, connection of portable lamps and appliances, elevator cables, wiring of cranes and hoists, connection of pieces of stationary equipment to facilitate their interchange, to prevent the transmission of noise or vibration, for fixed or stationary appliances where the fastening means and mechanical connections are designed to permit removal for maintenance and repair, for temporary wiring as allowed by *Sections 527.4(B)* and *527.4(C)* or for data processing cables as permitted by *Section 645.5* of the NEC (see NEC, *Section 400.7*).

11-2 What uses of flexible cord are prohibited?
Flexible cord must not be used as a substitute for the fixed wiring of a structure; run through holes in walls, floors, or ceilings; run through doors, windows, or similar openings; it must not be used for attaching to building surfaces or where it would be concealed behind building walls, ceilings, or floors (see NEC, *Section 400.8*).

11-3 Are splices or taps permitted in flexible cord?
No. Flexible cord may be used only in continuous length without splice or tap when initially installed in applications permitted by *Section 400.7(A)* of the NEC (see question 11-1). The repair of hard-service flexible cords size No. 14 and larger are permitted if conductors are spliced in accordance with *Section 110.14(B)* of the NEC, and the complete splice retains the insulation, outer sheath properties, flexibility, and usage characteristics of the cord being spliced (see NEC, *Section 400.9*).

11-4 What types of flexible cord may be used in show windows and showcases?
Types S, SO, SE, SEO, SEOO, SJ, SJE, SJO, SJEO, SJEOO, SJOO, ST, STO, STOO, SJT, SJTO, SJTOO, SO, SOO, SEW, SEOW, SEOOW, SJEW, SJEOW, SJEOOW, SJOW, SJOOW, SJTW, SJTOW, SJTOOW, SOW, SOOW, STW, STOW, or STOOW (see NEC, *Section 400.11*).

11-5 What is the minimum-size conductor permitted for flexible cord?

The individual conductors of a flexible cord or cable must be no smaller than the sizes shown in *Table 400.5* of the NEC.

11-6 What are the current-carrying capacities (in amperes) of various flexible cords?

Tables 400.5(A) and *400.5(B)* give the allowable current-carrying capacities for no more than three current-carrying conductors in a cord. If the number of current-carrying conductors in a cord is four to six, the allowable current-carrying capacity of each conductor will be reduced to 80 percent of the values in the table (see NEC, *Section 400.5*).

11-7 Does the equipment-grounding conductor of a flexible cord have to be identified?

Yes. It must be identified with a green color or continuous green color with one or more yellow stripes (see NEC, *Section 400.23*).

11-8 What is the minimum size allowed for fixture wires?

They may be no smaller than No. 18 wire (see NEC, *Section 402.6*).

11-9 What is the ampacity of a copper type W, single conductor flexible cord sized at No. 8 AWG?

60 amperes (see NEC, *Table 400.5(B)*).

11-10 What derating factors must be applied to a flexible cord that contains six current-carrying conductors?

80 percent of the value given in *Table 400.5* (see NEC, *Section 400.5*).

11-11 What does the term "ultimate insulation temperature" mean when applied to flexible cords and cables?

The limiting temperature of the conductors (see NEC, *Section 400.5*).

11-12 What is the ampacity of a 3-conductor type SEO No. 14 flexible cord?

15 amperes (see NEC, *Table 400.5(A)*).

11-13 Is an equipment-grounding conductor contained within a flexible cord or cable considered a current-carrying conductor?

Equipment for General Use **197**

No (see NEC, *Section 400.5*).

11-14 Are flexible cords and cables permitted for temporary wiring on construction sites?
Yes (see NEC, *Section 400.7(A)(11)*).

11-15 What is the allowable current-carrying capacity for fixture wires?
The ampacity of fixture wire must not exceed the following: (UTX 11.1)

Size (AWG)	Ampacity
18	6
16	8
14	17
12	23
10	28

No conductor should be used if its operating temperature will exceed the temperature specified in NEC, *Table 402.3* for the type of insulation involved.

11-16 What are the requirements, with respect to exposed live parts, on lighting fixtures?
There may be no live parts exposed, except on cleat receptacles and lampholders that are located at least 8 feet above the floor (see NEC, *Section 410.3*).

11-17 What are the requirements for fixtures located in damp areas?
The fixtures must be suitable for the location and must be installed so that moisture cannot enter the raceways, lampholders, or other electrical parts (see NEC, *Section 410.4*).

11-18 What precautions must be taken with fixtures installed near combustible materials?
They must be constructed, installed, or equipped with shades or guards so that combustible materials in the vicinity of the fixture won't be subject to temperatures in excess of 90°C (194°F) (see NEC, *Section 410.5*).

11-19 May externally wired fixtures be used in show windows?
No, with the exception of chain-supported fixtures (see NEC, *Section 410.7*).

11-20 Where must fixtures be installed in clothes closets?
A fixture in a clothes closet must be installed:

1. On the wall above the closet door, provided the clearance between the fixture and the storage area where combustible material may be stored within the closet is not less than 12 inches.

2. On the ceiling over an area that is unobstructed to the floor, maintaining a 12-inch clearance horizontally between the fixture and an area where combustible material may be stored within the closet.

   Note: A flush recessed fixture equipped with a solid lens is considered to be outside the closet area.

   You are permitted to install a flush recessed fixture that has a solid lens or a ceiling-mounted fluorescent fixture provided it is mounted or installed so that there is a minimum 6-inch clearance horizontally between the fixture and the storage area.

3. Pendants must not be installed in clothes closets (see NEC, *Section 410.8*).

11-21 How must flexible cords be connected to devices or fittings?
So that no tension will be transmitted to joints or terminals (see NEC, *Section 400.10*).

11-22 Must a conductor within a flexible cord be considered a current-carrying conductor when it serves as an equipment-grounding conductor and a grounding-neutral conductor for an electric range or electric clothes dryer?
No (see NEC, *Section 400.5*).

11-23 What must be done to flexible cords and cables where they pass through holes in covers, outlets boxes, or similar enclosures?
They require protection in the form of bushings or fittings (see NEC, *Section 400.14*).

11-24 Under what conditions may a light blue colored insulation be used for the identification of a grounded conductor in a flexible cord?
For jacketed cords furnished with an appliance (see NEC, *Section 400.22(C)*).

11-25 Under what conditions may multiconductor cables rated at over 600 volts, nominal, be used?

To connect mobile equipment and mobile machinery (see NEC, *Section 400.30*).

11-26 What precautions are necessary when installing fixture wires in temperatures colder than 10°C (14° F)?
Exercise care during installation, since the conductors may become brittle (see NEC, *Section 402.3*, FPN).

11-27 What is the ampacity of a No. 10 AWG fixture wire?
28 amperes (see NEC, *Table 402.5*).

11-28 How many fixture wires are permitted in a single conduit?
Not to exceed the capacity shown in Table 9-1 of Chapter 9 (see NEC, *Section 402.7*).

11-29 What is the maximum operating temperature of a Type TFFN fixture wire?
90°C (194°F) (see NEC, *Table 402.3*).

11-30 How are thermoplastic-insulated fixture wires marked?
On their surfaces at intervals not exceeding 24 inches (see NEC, *Section 402.9(A)*).

11-31 May fixture wire be used as a branch circuit conductor?
No (see NEC, *Section 402.11*).

11-32 What type of fixture is required for a corrosive location?
One that is suitable for such a location (see NEC, *Section 410.4(B)*).

11-33 May fixtures be installed in a nonresidential cooking hood?
Yes (see NEC, *Section 410.4(C)*).

11-34 What precautions must be considered when installing a pendant fixture above a bathtub?
It must be installed at least 8 feet from the top of the bathtub rim and 3 feet beyond the horizontal edge (see NEC, *Section 410.4(D)*).

11-35 What are the requirements for fixtures installed near combustible materials?
They must be constructed, installed, or equipped with shades or guards so that combustible materials won't be subjected to temperatures in excess of 90°C (194°F) (see NEC, *Section 410.5*).

11-36 Under what condition may a fixture be installed in a show window?

When it is chain-supported and externally wired (see NEC, *Section 410.7*).

11-37 Does the NEC contain a rule that is applicable to space for cove lighting?

Yes (see NEC, *Section 410.9*).

11-38 How many requirements must be met when installing metal poles used to support lighting fixtures?

Six (see NEC, *Section 410.15(B)*).

11-39 What are hickeys, tripods, and crowfeet?

Fittings used to support fixtures (see NEC, *Section 410.16(D)*).

11-40 How may fixtures be connected to busways?

In accordance with Section 368.12 (see NEC, *Section 410.16(G)*).

11-41 How are fixtures grounded?

By an equipment-grounding conductor (see NEC, *Section 410.21*).

11-42 Must fixtures be provided with a means for the connection of an equipment-grounding conductor?

Yes, when they have exposed metal parts (see NEC, *Section 410.20*).

11-43 What are the maximum operating temperatures and voltage limitations for fixture wires?

They must be in accordance with Section 402.3 (see NEC, *Section 410.24* FPN).

11-44 How must fixture conductors and insulation be protected?

In a manner that won't tend to cut or abrade the insulation (see NEC, *Section 410.28(A)*).

11-45 Are flexible cords permitted to connect fixed showcases in a department store?

No. (see NEC, *Section 410.29*).

11-46 When may a fixture be used as a raceway for circuit conductors?

When listed for use as a raceway (see NEC, *Section 410.31*).

11-47 May electric discharge lighting fixtures be connected by a cord and cap?

Yes, when the fixture is located directly below the outlet box, continuously visible, and not subject to strain (see NEC, *Section 410.30(C)*).

11-48 What must be tested before being connected to the circuit supplying a fixture?

All wiring must be tested and found to be free from short circuits and grounds before the fixture is connected to the circuit (see NEC, *Section 410.45*).

11-49 What are the requirements for the use of double-pole switched lampholders that are supplied by ungrounded circuit conductors?

The switching device must simultaneously disconnect both conductors of the circuit (see NEC, *Section 410.48*).

11-50 What marking is required for receptacles rated at 20 amperes or less that are directly connected to aluminum conductors?

Co/ALR (see NEC, *Section 406.2(C)*).

11-51 May a receptacle outlet be installed in a shower space?

No (see NEC, *Section 406.8(C)*).

11-52 What type of wire must be used in the connection of fixtures?

The wire must have the type of insulation that will withstand the temperatures to which the fixture will be exposed. Check the temperature rating of the wire against the temperature at which the fixture operates to obtain the type of insulation required (see NEC, *Section 410.11*).

11-53 May branch-circuit wiring pass through an outlet box that is an integral part of an incandescent fixture?

No, unless the fixture is identified for the purpose (see NEC, *Section 410.11*).

11-54 When a fixture weighs more than 50 pounds, may it be supported from the outlet box?

It must be supported independently of the outlet box (see NEC, *Section 314.27(B)*).

11-55 What is the minimum-size conductor allowed for fixture wires?

No. 18 wire (see NEC, *Section 402.6*).

11-56 What factors must be considered when choosing fixture wires?
Operating temperature, corrosion and moisture conditions, voltage, and current-carrying capacity (see NEC, *Section 410.24*).

11-57 What precautions must be taken in conductors for movable parts of fixtures?
Stranded conductors must be used and must be arranged so that the weight of the fixture or movable parts won't put tension on the conductors. These measures must be taken to protect the conductors (see NEC, *Section 410.28(F)*).

11-58 What types of insulation are permitted within 3 inches of a ballast within a ballast compartment?
Conductors rated at 90°C (194°F) (see NEC, *Section 410.3331*).

11-59 Which wire must be connected to the screw shells of lampholders?
The grounded, or white, conductor (see NEC, *Section 410.47*).

11-60 What is the maximum wattage permitted on a medium lamp base?
300 watts (see NEC, *Section 410.53*).

11-61 What is the maximum lamp wattage permitted for a mogul lamp base?
1500 watts (see NEC, *Section 410.53*).

11-62 For flush or recessed fixtures, what is the maximum operating temperature permitted when installed in or near combustible materials?
90°C (194°F) (see NEC, *Section 410.65*).

11-63 What precautions must be taken when switching off discharge-lighting systems that are rated at 1000 volts or more?
The switch must be located in sight of the fixtures or lamps, or it may be located elsewhere, provided it is capable of being locked in the open position (see NEC, *Section 410.81(B)*).

11-64 When fixtures are mounted in suspended ceilings, what is required?
Framing members of the suspended ceilings must be securely fastened together, and the fixtures must be securely fastened to

the ceiling frames by bolting, screws, rivets, or by clips identified for the purpose (see NEC, *Section 410.16(C)*).

11-65 How must fixtures be grounded?

They may be connected to metal raceways, the armor of metal-clad cable, etc., if properly installed and grounded, or a separate equipment-grounding conductor not smaller than No. 14 wire may be used (see NEC, *Sections 410.21, 250.118, and 250.122*).

11-66 How must immersion-type portable heaters be constructed?

They must be constructed and installed so that current-carrying parts are effectively insulated from the substance in which they are immersed (see NEC, *Section 422.44*).

11-67 Must electric flatirons have temperature-limiting devices?

Yes (see NEC, *Section 422.46*).

11-68 What type of disconnection means must be provided on stationary appliances?

On appliances rated at less than 300 volt-amperes or ⅛ hp, the branch-circuit overcurrent-protective device may serve as the disconnecting means. For stationary appliances of greater rating, the branch-circuit switch or circuit breaker may serve as the disconnecting means, provided that it is readily accessible to the user of the appliance. For ranges, dryers, and any other cord-connected appliance, the plug at the receptacle may suffice as the disconnecting means (see NEC, *Section 422.31*).

11-69 On motor-driven appliances, what disconnecting means must be provided, and how must it be located?

The switch or circuit breaker that serves as the disconnecting means on a stationary motor-driven appliance must be located within sight of the motor controller or be capable of being locked in the open position (see NEC, *Section 422.31(B)*).

11-70 On all space-heating systems, what is the main requirement?

All heating equipment must be installed in an approved manner (see NEC, *Section 424.9*).

11-71 What are the requirements for heating cables?

Heating cables must be supplied complete with factory-assembled nonheating leads of at least 7 feet in length (see NEC, *Section 424.34*).

11-72 What markings must be present on heating cables?
Each unit length must have a permanent marking located within 3 inches of the terminal end of the nonheating leads, with the manufacturer's name or identification symbol, catalog number, and the rating in volts and watts or amperes. The leads on a 240-volt cable must be red; on a 120-volt cable the leads are yellow; 208-volt cable leads are blue; and on a 277-volt cable the leads are brown (see NEC, *Section 424.35*).

11-73 With fixed electric space-heating, must controllers or disconnecting means open all ungrounded conductors?
Devices that have an "off" position must open all ungrounded conductors; thermostats without an "on" or "off" position don't have to open all ungrounded conductors or remote control circuits (see NEC, *Section 424.20*).

11-74 What clearance must be given to wiring in ceilings?
Wiring above heated ceilings and contained within thermal insulation must be spaced not less than 2 inches above the heated ceiling and will be considered as operating at 50°C. Wiring above heated ceilings and located over thermal insulation having a minimum thickness of 2 inches requires no correction for temperature. Wiring located within a joist space having no thermal insulation must be spaced not less than 2 inches above the heated ceiling and will be considered as operating at 50° (see NEC, *Section 424.36*).

11-75 What clearance is required for wiring in walls where electrical heating is used?
Wires on exterior walls must comply with Article 300 and Section 310-10 (see NEC, *Section 424.37*).

11-76 What are the restricted areas for heating?
Heating panels must not extend beyond the room or area in which they originate; cables must not be installed in closets, over cabinets that extend to the ceiling, under walls or partitions or over walls or partitions that extend to the ceiling (see NEC, *Section 424.38*).

11-77 If heating is required in closets for humidity control, may it be used?
Yes, low-temperature heating sources may be used (see NEC, *Section 424.38(C)*).

11-78 What clearance must be provided for heating cables and panels from fixtures, boxes, and openings?

Panels and heating cables must be separated by a distance of at least 8 inches from lighting fixtures, outlets, and junction boxes, and 2 inches from ventilation openings and other such openings in room surfaces, or at least a sufficient area must be provided (see NEC, *Section 424.39*).

11-79 May embedded cables be spliced?

Only when absolutely necessary, and then only by approved means; in no case may the length of the cable be altered (see NEC, *Section 424.40*).

11-80 May heating cable be installed in walls?

No (see NEC, *Section 424.41(A)*).

11-81 What is the spacing requirement on heating cable in dry wall and plaster?

Adjacent runs of heating cable not exceeding 2¾ watts per foot must be installed not less than 1½ inches on centers (see NEC, *Section 424.41(B)*).

11-82 How is heating cable installed in dry board and plaster?

Heating cables may be applied only to gypsum board, plaster lath, or other similar fire-resistant materials. On metal lath or other conducting surfaces, a coat of plaster must be applied first. The entire ceiling must have a coating of ½ inch or more of thermally noninsulating sand plaster or other approved material. Cable must be secured at intervals not exceeding 16 inches with tape, staples, or other approved devices; staples may not be used with metal lath. On dry-board installations, the entire ceiling must be covered with gypsum board not exceeding ½ inch in thickness, and the void between the upper and lower layers of gypsum board must be filled with thermally conducting plaster or other approved material (see NEC, *Section 424.41(G)*).

11-83 May the excess of the nonheating leads on heating cable be cut off?

No, they must be secured to the under side of the ceiling and embedded in the plaster. The ends have a color-coding on them, and these must be visible in the junction box (see NEC, *Section 424.43(C)*).

11-84 How is heating cable installed in concrete?
Panels or heating units shall not exceed 16½ watts per linear foot of cable. Adjacent runs of cable shall be a minimum of 1 inch apart. The cable shall be secured with frames or spreaders (non-metallic) while the concrete is being applied. A minimum space of 1 inch must be maintained between the cable and other metallic objects. The leads extending from the concrete shall be protected by rigid metal conduit, intermediate metal conduit, rigid non-metallic conduit, or EMT. Bushings shall be used where cable leads emerge from the floor slab (see NEC, *Section 424.44*).

11-85 What tests must be run on cables during and after installation?
They must be tested to verify that they have not been damaged. Generally this requires testing for continuity and insulation resistance (see NEC, *Section 424.45*).

11-86 What percentage of branch-circuit ratings is allowable for use with air-conditioning units?
When supplying only the air-conditioning unit, not more than 80 percent of the circuit rating may be used. When lighting or other appliances are also supplied on the same branch circuit, not more than 50 percent of the circuit rating may be used for the air-conditioning unit (see NEC, *Section 440.62(B)* and *440.62(C)*).

11-87 May a plug and receptacle be used as the disconnecting means for an air-conditioning unit?
An attachment plug and receptacle may serve as the disconnecting means for a single-phase, room air-conditioning unit, rated 250 volts or less, when

1. The manual controls on the air-conditioning units are readily accessible and located within 6 feet of the floor.

2. An approved manually operated switch is installed in a readily accessible location within sight of the air-conditioning unit (see NEC, *Section 440.63*).

11-88 What is a sealed (hermetic-type) motor compressor?
This is a mechanical compressor and a motor, enclosed in the same housing, with neither an external shaft nor seal; the motor operates in the refrigerant atmosphere (see NEC, *Section 440.2*).

11-89 What does "in sight from" mean as it is applied to motors?
When "in sight from" is used with reference to some equipment in relation to other equipment, the term means that the

equipment must be visible and within 50 feet of each other (see NEC, *Article 100*, for definition).

11-90 How many parts does a motor feeder and branch circuit contain?
Up to nine (see NEC, *Diagram 430.1*).

11-91 What is the purpose of a locked-rotor table?
A locked-rotor table indicates the current when full voltage is applied to a motor, with the rotor held in a locked position (see NEC, *Tables 430.151(A)* and *430.151(B)*).

11-92 How must hermetic-type refrigeration compressors be marked?
They must have a nameplate, giving the manufacturer's name, the phase, voltage, frequency, and full-load current in amperes of the motor current when the compressor is delivering the rated output. If the motor has a protective device, the nameplate must be marked "Thermal Protection." For complete details, see *Section 440.4(A)* of the NEC.

11-93 What markings must be provided on motor controllers?
Maker's name or identification, the voltage, and the current or horsepower rating (see NEC, *Section 430.8*).

11-94 Must the controller be the exact size for the motor it is to be used on?
No, this is not necessary; however, it must be at least as large as is necessary for the job it has to do. It may be larger, but the overloads that might be used in conjunction with it should be the proper size for the application (see NEC, *Section 430.83*).

11-95 When using switches, fuses, and other disconnecting means, must the voltage of the switch coincide with that of the motor?
Yes. The voltage rating of the switch may not be less than 115 percent of the full load current rating of the motor (see NEC, *Section 430.110*).

11-96 If a switch is used that has a current rating larger than is required, may it be adapted for less current?
Yes, adapters of the approved type are available for reducing the fuse size of the switch; fuses larger than the switch rating, however, may not be used.

11-97 May the enclosures for controllers and disconnecting means for motors be used as junction boxes?

Enclosures for controllers and disconnecting means for motors may not be used as junction boxes, auxiliary gutters, or raceways for conductors feeding through or tapping off to the other apparatus unless designs are employed that provide adequate space for this purpose (see NEC, *Section 430.10(A)*).

11-98 If a branch circuit supplies only one motor, how large must the conductor size of the branch circuit be in comparison to the current rating of the motor?

The branch-circuit conductors should be figured at not less than 125 percent of the full-load current rating of the motor (see NEC, *Section 430.22(A)*).

11-99 What are the torque requirements for control circuit devices with screw-type pressure terminals with No. 14 AWG or smaller copper conductors?

They must be torqued to a minimum of 7 pound inches (see NEC, *Section 430.9(C)*).

11-100 What type of conductor material is required for connections to motor controllers and terminals of motor control circuit devices?

Copper conductors, unless identified for use with a different conductor (see NEC, *Section 430.9(B)*).

11-101 What are the conductor rating factors for power resistors that operate for a continuous duty constant voltage dc motor?

110 percent of the ampacity of conductors (see NEC, *Section 430.29* and *Table 430.29*).

11-102 What is a motor control circuit?

See definition in NEC, *Section 430.71*.

11-103 What is the maximum rating of an overcurrent device in amperes for a control circuit conductor sized at No. 18 AWG?

7 amperes (see NEC, *Section 430.72(B)* and *Table 430.72(B)*).

11-104 What size wire must be used to connect the wound-motor secondary of a motor?

For continuous duty, the conductor size must not be less than 125 percent of the full-load secondary current of the motor (see NEC, *Section 430.23*).

11-105 When conductors supply several motors, how can you determine the current-carrying capacity of the conductors?

Conductors may not use less than 125 percent of the full-load current rating of the highest-rated motor in the group and not less than the full-load current rating of the rest of the motors. The total current is the basis for calculating the wire size (see NEC, *Section 430.24*).

11-106 What overload-current protection must motors of more than 1 hp have?

Motors having a temperature rise of not more than 40°C must trip at not more than 125 percent of full-load current (see NEC, *Section 430.32*).

11-107 How many conductors must be opened by motor overcurrent devices?

A sufficient number of underground conductors must be opened to stop the motor (see NEC, *Section 430.38*).

11-108 What are the number and location of overcurrent-protection units for different motors?

See NEC, *Table 430.37*.

11-109 How many overcurrent-protective devices are required on a three-phase motor?

One in each phase unless protected by other approved means (see NEC, *Table 430.37*, Exception).

11-110 Where the motor and driven machinery are not "in sight from" the controller location, what conditions must be met?

The controller disconnection means must be capable of being locked in the open position, or there must be, within sight of the motor location, a manually operated switch that will disconnect the motor from its source of supply (see NEC, *Section 430.102*).

11-111 How many motors may be served from one controller?

Each motor must have its own controller; however, where the motors are rated at 600 volts or less and drive several parts of one machine, are under the protection of one overcurrent-protection device, or where a group of motors is located in one room "in sight from" the controller, a single controller may serve a group of motors (see NEC, *Section 430.87*).

11-112 Must the disconnecting means be marked to indicate the position it is in?
Yes, it must be plainly marked to indicate whether it is in the open or closed position (see NEC, *Section 430.104*).

11-113 May the service switch be used as the disconnecting means?
Yes, if it is "within sight" from the controller location, or if it is capable of being locked in the open position (see NEC, *Section 430.109*).

11-114 Must the disconnecting means be readily accessible?
Yes (see NEC, *Section 430.107*).

11-115 What should the current-carrying capacity of the disconnecting means be?
It must have a current-carrying capacity of at least 115 percent of the full-load current rating of the motor (see NEC, *Section 430.110*).

11-116 A single-phase motor draws 32 amperes under full-load conditions. How large must the fuses and branch-circuit switch be?
They must be rated at 300 percent of the full-load current, or about 100 amperes each (see NEC, *Table 430.150*).

11-117 In a group of four motors, one motor draws 10 amperes, one draws 45 amperes, and two draw 75 amperes. What size conductors must be used for the feeder circuit?

$$75 \text{ amperes} \times 125\% = 94 \text{ amperes}$$
$$94 + 75 + 45 + 10 = 224 \text{ amperes}$$

The feeder conductors, therefore, must be able to handle 224 amperes (see NEC, *Section 430.24*).

11-118 What is the full-load current for a synchronous type 75 hp motor operating at 3-phase 460 volts?
78 amperes (see NEC, *Table 430.150*).

11-119 What is the locked rotor current in amperes for a 10 hp, type C motor operating at 3-phase 230 volts?
179 amperes (see NEC, *Table 430.151(B)*).

11-120 Motor locked-rotor currents are approximately how many times the full-load current values given in *Tables 430.150* and *430.151(B)*?
Approximately six times.

11-121 What is the full-load current of a 3 hp motor operating on a single-phase 115-volt system?
    34 amperes (see NEC, *Table 430.148*).

11-122 What is the maximum rating or setting of a motor branch-circuit short-circuit and ground fault protective device for a single-phase motor with no code letter?
    The percent of the full-load current of the motor at 300 percent for a nontime delay fuse; 175 percent for a dual element (time delay) fuse; 700 percent for an instantaneous trip circuit breaker; and 250 percent for an inverse time circuit breaker (see NEC, *Table 430.52*).

11-123 What is the maximum percentage of a full-load current permitted for a high reactance squirrel cage motor rated at not more than 30 amperes with no code letter, using an instantaneous trip circuit breaker?
    700 percent of full load current of the motor (see NEC, *Table 430.52*).

11-124 What are code letters and how are they explained?
    See NEC, *Section 430.7(B)* and *Table 430.7(B)* for code letters and their limitations.

11-125 For small motors not covered in NEC *Tables 430.147, 430.148, 430.149, or 430.150,* what is the locked-rotor current?
    They are assumed to be six times the full-load current of the motor (see NEC, *Section 430.110(C)(3)*).

11-126 What is the full-load current for a two-phase 4-wire ac motor sized at 60 hp and operating at 230 volts?
    133 (see NEC, *Table 430.149*).

11-127 What is the full-load current for a 3-phase ac wound-rotor 100 hp motor operating at a voltage of 460 volts?
    124 amperes (see NEC, *Table 430.150*).

11-128 What is the full-load current for a 3-phase ac squirrel cage 50 hp motor operating at 575 volts?
    52 (see NEC, *Table 430.150*).

11-129 Must a controller enclosure have a means for the attachment of an equipment-grounding conductor?
    Yes (see NEC, *Section 430.144*).

11-130 What is the full-load current in amperes for a 10 hp dc motor operating at 120 volts?

 76 amperes (see NEC, *Table 430.147*).

11-131 What is full-load current in amperes for a single-phase ac motor sized at 1 hp and operating at 115 volts?

16 amperes (see NEC, *Table 430.148*).

11-132 What is a "controller" when it operates to control a motor?

It is any switch or device that is normally used to stop and start a motor by making and breaking the motor circuit current (see NEC, *Section 430.81*).

11-133 What should the horsepower rating of a controller be?

Not less than the horsepower rating of the motor (see NEC, *Section 430.83(A)*).

11-134 What type of motor controller is used when its submersion is prolonged occasionally?

Enclosure type 6P (see NEC, *Section 430.91* and *Table 430.91*).

11-135 Can a motor controller serve as an overload device?

Yes, when the number of overload units complies with NEC, *Section 430.39*.

11-136 What are the requirements for fuses used as overload protection for a motor?

A fuse must be inserted in each ungrounded conductor (see NEC, *Section 430.36*)

Chapter 12

Motors

12-1   What is a dc motor?
This is a motor designed and intended for use with direct current only.

12-2   If a dc motor is connected across an ac source, what will be the result? Why?
Direct-current flow is obstructed only by resistance, whereas alternating current is obstructed by both resistance and inductive reactance. Therefore, when a dc motor is connected across an ac source, the current on ac will be much less than that of dc. The motor would run, however, but it would not carry the same load as it would on dc. There would be more sparking at the brushes. The armature is made up of laminations, but the field is not. The eddy currents in the field would therefore cause the motor to heat up and eventually burn out on ac; this would not happen on dc.

12-3   What is the difference between a dc motor and a dc generator?
Fundamentally there is none. A dc motor will generate electricity if driven by some prime mover, and a dc generator will act as a motor if connected across a dc source.

12-4   What are the three fundamental types of dc motors?
They are the series-wound motors, the shunt-wound motors, and the compound-wound motors.

12-5   What is a series-wound motor? Draw a schematic of one.
In a series-wound motor, the field is wound in series with the armature (Figure 12-1).

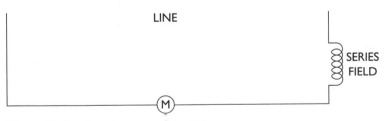

Figure 12-1   A series-wound motor.

12-6 What is a shunt-wound motor? Draw a schematic of one.

In a shunt-wound motor the field winding is in shunt, or parallel, with the armature. The field consists of many turns of small wire since it must be able to handle the line voltage that is connected across it (Figure 12-2).

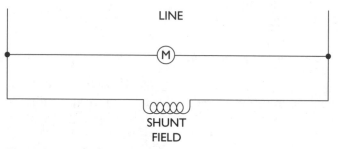

Figure 12-2 A shunt-wound motor.

12-7 What is a compound-wound motor? Draw a schematic of one.

A compound-wound motor is a combination of a series- and a shunt-wound motor and has better speed regulation than either one (Figure 12-3).

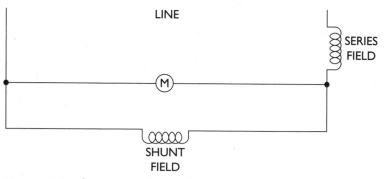

Figure 12-3 A compound-wound motor.

12-8 If the field on a dc motor were opened, what would happen?

The motor would try to run away with itself, or, in other words, the motor would reach a very high speed and might destroy itself.

12-9   What are the advantages and disadvantages of dc motors compared with ac motors?

Speed control of dc motors is much easier, making them more versatile for use where a wide range of speeds is required. They may be used for dynamic braking; that is, a motor on an electric train will act as a motor when required, but when going downhill, it can be used as a generator, thereby putting current back into the line. In generating, it requires power, so it acts as a brake. Dc motors, however, require more maintenance than most ac motors. In order to use a dc motor, special provisions must be made.

12-10   What is a motor-generator set?

This is a generator driven by a motor, with both devices mounted on a common base. It might be an ac motor driving a dc generator or a dc motor driving an alternator.

12-11   What is a rotary converter?

This is a self-contained unit having two or more armature windings and a common set of field poles. One armature winding receives the direct current and rotates, thus acting like a motor, whereas the others generate the required voltage, and thus act as generators.

12-12   If a dc motor is rapidly disconnected from the line, what must be provided?

A means of shorting out the motor, either directly or through a resistance, so that the collapsing magnetic field does not induce a high voltage.

12-13   What different types of alternating current are used on motors?

Single-phase, two-phase, three-phase, and six-phase; 25-cycle, 50-cycle, 60-cycle, and other frequencies may also be used.

12-14   May single-phase motors be run on two- or three-phase lines?

Yes, if they are connected to only two phase wires.

12-15   Can a three-phase motor be run on a single-phase line?

Yes, but a phase splitter must be used.

12-16   What is a phase splitter?

This is a device, usually composed of a number of capacitors connected in the motor circuit, that produces, from a single input wave, two or more output waves that differ in phase from each other.

12-17 When using a phase splitter, there will be some strange current readings. What will they be?

At no-load conditions, the current on the three motor leads will be unbalanced: One will have a high current, and the other two will have a low current. As the motor is loaded, these currents will begin to balance out, and at full-load conditions they will have equalized.

12-18 Why would you use a phase splitter and a three-phase motor instead of a single-phase motor?

It is possible that at the moment single-phase power is all that is available, but in the future three-phase power is expected. Therefore, if you purchase the three-phase motor and a phase splitter, the wiring will be in place, the motor will be at the desired location, and the expenses will be cut down. There is also less maintenance on three-phase motors; this one fact will often influence the use of a phase splitter.

12-19 What causes a motor to turn?

There are two parts to a motor: a stator, or field, and a rotor, or armature. Around one part there exists a magnetic field from the line current, and in the other part there is an induced current that causes a magnetic field of opposite polarity. These magnetic fields repel one another, thereby causing the rotating member to turn.

12-20 What is termed the "front end" of a motor? Draw a sketch.

The end opposite the shaft is termed the "front end" (Figure 12-4).

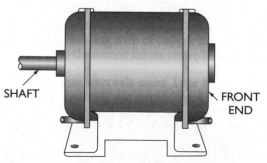

SHAFT

FRONT END

Figure 12-4 The "front end" of a motor.

12-21 What is the standard direction of rotation for a motor?
Counterclockwise—that is, the shaft of the motor appears to be turning counterclockwise when you are looking at the front end of the motor.

12-22 Illustrate the standard and opposite directions of rotation of a motor.
See Figure 12-5.

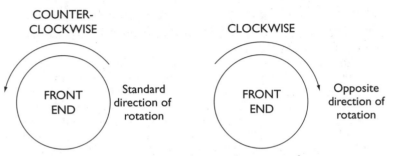

Figure 12-5 The standard and opposite directions of rotation of a motor.

12-23 What are several types of three-phase motors?
Squirrel cage, wound-rotor, and synchronous.

12-24 What is a squirrel cage motor?
This is a motor with the winding on the stator, or line winding. The rotor consists of a winding made of bars that are permanently short-circuited at both ends by a ring.

12-25 How are the voltage and current produced in a squirrel cage rotor?
The stator can be considered the primary winding, and the rotor the short-circuited secondary winding of a transformer. Thus, the voltage and current are induced in the rotor.

12-26 How is regulation obtained in a motor?
The motor at start is similar to a transformer with a shorted secondary. The current in the rotor and stator will be high. As the motor approaches its rated speed, the rotor induces a voltage into the stator in opposition to the line voltage; this is called *counter emf.* The line current is then reduced in proportion to the speed.

12-27 Does an ac motor (other than a synchronous motor) run at synchronous speed?
No. It must slip below synchronous speed so that an effective voltage will be produced.

12-28 What is a synchronous motor?
A synchronous motor is almost exactly the same as an alternator. The field must be excited by direct current. The motor runs at the same speed or at a fixed multiple of the speed of the alternator supplying the current for its operation. Should it slip, the motor will pull out and stop since it must run pole for pole with the alternator.

12-29 How do synchronous motors differ from alternators?
They may be just like alternators; however, if they are, they won't be self-starting and will have to be started by some means until they approach synchronous speed, at which time they can be connected to the line and pull into speed. Most synchronous motors have a squirrel cage winding in addition to the dc field. They start as a squirrel cage motor, and when they are about up to the speed of the alternator, the dc field is energized. The poles then lock in position with the revolving field of the armature, and the rotor revolves in synchronization with the supply circuit.

12-30 If the field of a synchronous motor is underexcited, what will happen to its power factor?
It will lag.

12-31 If the field of a synchronous motor is overexcited, what will happen to the power factor?
It will lead.

12-32 What is a synchronous capacitor?
It is a synchronous motor running without mechanical load on the line, with its field overexcited for power-factor correction.

12-33 In what direction does the rotor turn in an alternating-current motor?
It always turns in the direction of the rotating field (Figure 12-6).

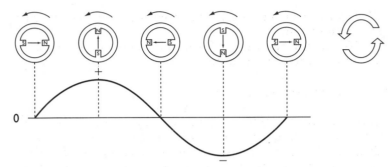

Figure 12-6 The direction of rotor rotation in an ac motor.

12-34 On three-phase motors, how may the fields, or stator windings, be connected internally?
They may be connected in either a delta or a wye arrangement.

12-35 Draw a schematic of a squirrel cage motor connected in a delta arrangement.
See Figure 12-7.

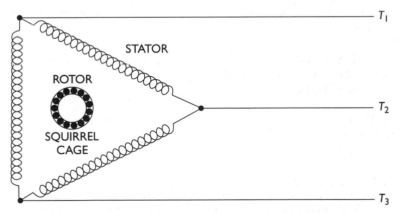

Figure 12-7 A squirrel cage motor connected in a delta arrangement.

12-36 Draw a diagram of a delta-wound motor. Number the coils and show how they must be connected for a motor that can be used on a 240/480-volt system.
See Figure 12-8.

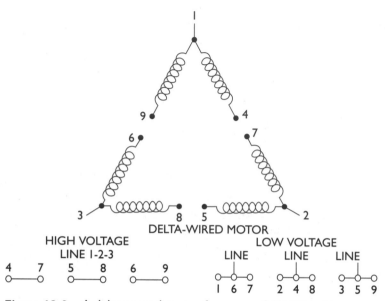

Figure 12-8 A delta-wound motor for use on 240/480 volts.

12-37 How are motor leads numbered?
T_1, T_2, T_3, T_4, etc.

12-38 Draw a diagram of a wye-wound motor. Number the coils and show how they must be connected for a motor that can be used on 240/480 volts.
See Figure 12-9.

12-39 What is an easy method for remembering which leads to connect together for a wye-wound motor that has two voltage ratings?
First, draw the windings in a wye arrangement. Then, starting at one point, draw a spiral so that it connects in order with all six coils. Begin at the starting point and number from 1 to 9 as you progress around the spiral (Figure 12-10).

12-40 What are the different kinds of single-phase motors?
Split-phase, capacitor, capacitor-start, capacitor-start-capacitor-run, shading-pole, repulsion, repulsion-induction, and repulsion-start-induction-run.

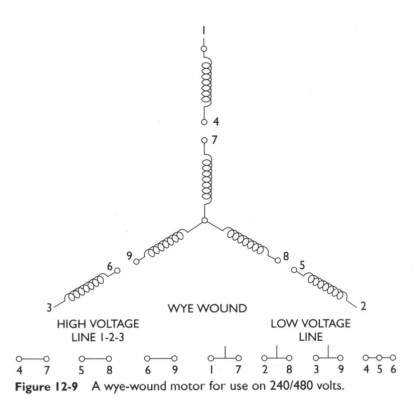

Figure 12-9 A wye-wound motor for use on 240/480 volts.

12-41 What is a wound-rotor, three-phase motor?

This is a three-phase motor that has another three-phase winding instead of a squirrel cage rotor, the terminals of which are connected to three slip rings. Brushes ride these slip rings and deliver the current to an external three-phase rheostat or variable resistor. At start, all the resistance is in the circuit; as the motor picks up speed, the resistance is gradually decreased until finally the slip rings are short-circuited.

12-42 Can a wound-rotor motor be used as a variable-speed motor?

Yes, if it has been properly designed for this purpose.

12-43 How can you reverse the direction of a three-phase squirrel cage motor?

By transposing any two of the motor leads.

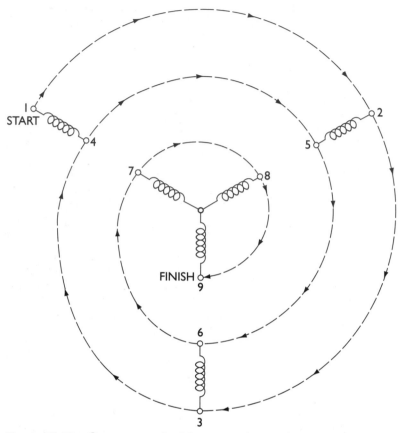

Figure 12-10   One easy method for remembering the proper connections for a wye-wound motor with two voltage ratings.

12-44 How can you change the rotation of a wound-rotor motor?
By transposing any two of the line leads.

12-45 Can a wound-rotor motor be reversed by transposing any two leads from the slip rings?
No. There is only one way to reverse its direction and that is by transposing any two line leads.

12-46 Draw a schematic diagram of a wound-rotor motor.
See Figure 12-11.

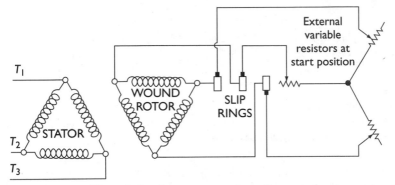

Figure 12-11   The schematic representation of a wound-rotor motor.

12-47 What is a split-phase motor?

This is a single-phase induction motor equipped with an auxiliary winding connected in shunt with the main stator winding that differs from it in both phase and spacing. The auxiliary winding is usually opened by a centrifugal device when the motor has reached a predetermined speed. The fields of these two windings act on the rotor to produce a small starting torque. Once the motor is started, it produces its own rotating field and no longer requires the starting torque.

12-48 Does a split-phase motor have a high or low starting torque?

It has a low starting torque.

12-49 How can you reverse the rotation of a split-phase motor?

By reversing the leads to the running or the starting winding, but not both.

12-50 Draw two schematic diagrams illustrating the reversal of a split-phase motor by reversing the running winding leads.

See Figure 12-12.

12-51 What is a capacitor-start motor?

Fundamentally, this motor is very similar to the split-phase motor except that the starting winding has a few more turns and consists of heavier wire than the starting winding of a split-phase motor. There is also a large electrolytic capacitor connected in series with the starting winding. The capacitor and starting winding are cut out of the circuit as soon as the motor reaches approximately 70 percent of its full speed.

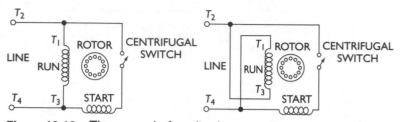

Figure 12-12 The reversal of a split-phase motor is accomplished by reversing the running winding leads.

12-52 Does the capacitor-start motor have a high or low starting torque?

It has a high starting torque. For this reason and the fact that it requires little maintenance, it is quickly replacing the repulsion types of motors; it is also cheaper to build, and the rewinding is more practical than the repulsion-type motors.

12-53 How can you reverse the direction of rotation of a capacitor-start motor?

By reversing either the running- or starting-coil leads where they are connected to the line. Don't reverse both.

12-54 Are capacitor-start motors usually dual-voltage or single-voltage motors?

They are usually dual-voltage motors.

12-55 If a capacitor-start motor is rated as a 115/230-volt motor (the capacitor is usually a 115-volt capacitor), how is it connected, when the motor is used on 230 volts, to protect the capacitor from being damaged?

The running winding is in two sections so that it may be connected in parallel for 115 volts and connected in series for 230 volts. The capacitor is in series with the starting winding; therefore, when operating at 115 volts, the winding leads are connected across the line at start, and when operating at 230 volts, one lead is connected to one side of the line and the other lead is connected to the midpoint of the running windings.

12-56 Draw two schematic diagrams of a capacitor-start motor, with one showing the 115-volt connection and one showing the 230-volt connection.

See Figure 12-13.

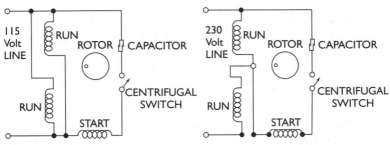

Figure 12-13 A capacitor-start motor, first connected across 115 volts and then across 230 volts.

12-57 What is a capacitor-start-capacitor-run motor?
This is the same as the capacitor-start motor except that it has an extra capacitor (oil type) connected in the starting winding, which is always in the motor circuit. The starting capacitor is in the starting circuit only when the motor is started and is disconnected by means of a centrifugal switch.

12-58 What are the advantages of a capacitor-start-capacitor-run motor over the capacitor-start motor?
It is smoother running, has a higher power factor, and consequently draws less current than the capacitor-start motor.

12-59 How can the direction of a capacitor-start-capacitor-run motor be changed?
By reversing the leads to either the starting winding or the running winding, but not both.

12-60 Draw a schematic diagram of a capacitor-start-capacitor-run motor.
See Figure 12-14.

12-61 Draw a schematic diagram showing the direction reversal of a capacitor-start motor.
See Figure 12-15.

12-62 Draw a schematic diagram of a two-speed capacitor motor.
See Figure 12-16.

12-63 What is a shaded-pole motor?
This is a single-phase induction motor that is provided with one or more auxiliary short-circuited stator windings that are magnetically displaced from the main winding.

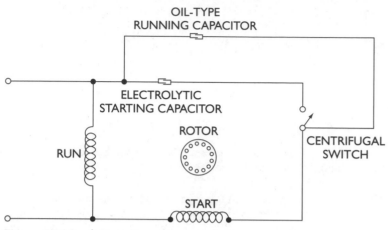

Figure 12-14   A capacitor-start-capacitor-run motor.

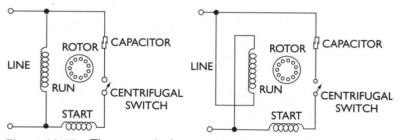

Figure 12-15   The reversal of a capacitor-start motor is accomplished by reversing the leads of the running winding.

12-64 Does a shaded-pole motor have much starting torque?
        No, it has very little starting torque, and for this reason it is used primarily for small fan motors.

12-65 In what direction do shaded-pole motors run?
        They rotate toward the shading coil.

12-66 Draw a sketch of one pole of a shaded-pole motor showing the direction of rotation.
        See Figure 12-17.

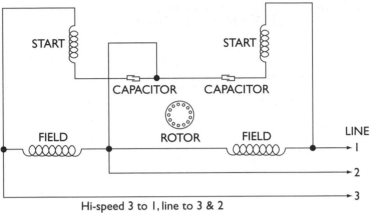

Hi-speed 3 to 1, line to 3 & 2
Lo-speed 1 & 3 to line, 2 no connection

Figure 12-16 The schematic representation of a two-speed capacitor motor.

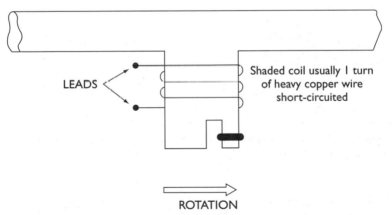

Figure 12-17 One pole of a shaded-pole motor.

12-67 What is a repulsion motor?

This motor has a stator like that of most single-phase motors and a rotor similar to that of a dc motor. It has a field and an armature; however, the armature is not connected to the supply source, and the brushes are short-circuited, or connected together, by a conductor of negligible resistance.

12-68 Draw a schematic diagram of a repulsion motor.
See Figure 12-18.

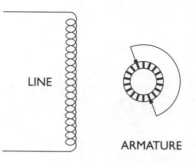

Figure 12-18   The schematic representation of a repulsion motor.

LINE

ARMATURE

12-69   What is a repulsion-induction motor?

This is a repulsion motor with a squirrel cage rotor on the armature, so that after the motor starts, it will run as an induction motor.

12-70   What is a repulsion-start-induction-run motor?

This is a repulsion motor that has a centrifugal-force device to short-circuit the commutator when the motor reaches its rated speed.

12-71   How can the direction of rotation for the repulsion, repulsion-induction, and repulsion-start-induction-run motors be changed?

By shifting the position of the brushes about 15° (electrical).

12-72   What is a universal motor?

This is a motor built like a series dc motor. However, the stator and armature are both laminated, designed for high speeds, and may be used on either ac or dc, although the speed and power will be greater on dc.

12-73   How can the direction of rotation of a universal motor be changed?

By reversing either the field leads or the armature leads, but not both.

12-74   How can the direction of rotation of a dc motor be changed?

By reversing either the field leads or the armature leads, but not both.

12-75 Knowing the frequency and number of poles of an ac motor, how can its speed be determined? This will be synchronous speed; the actual speed will be slightly lower.

The speed of an ac motor can be found by using the formula:

$$\text{Speed} = \frac{120 \times \text{frequency}}{\text{number of poles}}$$

12-76 We have a 60-cycle, 4-pole ac motor. What will its speed be?

$$\text{Speed} = \frac{120 \times 60}{4} = \frac{7200}{4} = 1800 \text{ rpm}$$

Chapter 13

Motor Controls

Motor controls are not really complicated. Once you master the fundamental idea, you will be able to figure out and sketch all types of controls. First, you must become familiar with the common symbols (see Figure 13-1) that are used in connection with controls; second, analyze what you want to accomplish with the particular control. If you follow through, your diagram will fit right into place.

There are two common illustrative methods—*diagrams* and *schematics*. Both are used, and in practically every case either one will be acceptable to the examiner. However, you should ask the examiner to be certain that you may use either; this gives you the opportunity to use the one you are most familiar with. The schematic method, however, presents a clearer picture of what you are trying to accomplish and is easier for the examiner to follow and correct. Whichever method you use, take your time and make a clear sketch. There is nothing as irritating to an examiner as trying to determine what your intent was and whether you had any idea of what you were doing.

Common Symbols for Motor Controls

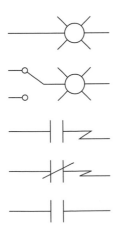

PILOT LAMP

PILOT LAMP WITH PUSHBUTTON TO TEST

NORMALLY OPEN CONTACTOR WITH BLOWOUT

NORMALLY CLOSED CONTACTOR WITH BLOWOUT

NORMALLY OPEN CONTACTOR

(continued)

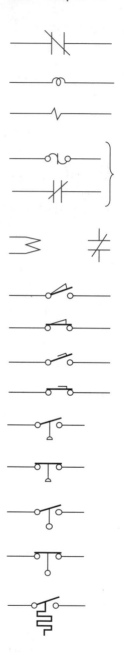

NORMALLY CLOSED CONTACTOR

SHUNT COIL

SERIES COIL

THERMAL OVERLOAD RELAY

MAGNETIC RELAY

LIMIT SWITCH, NORMALLY OPEN

LIMIT SWITCH, NORMALLY CLOSED

FOOT SWITCH, NORMALLY OPEN

FOOT SWITCH, NORMALLY CLOSED

VACUUM SWITCH, NORMALLY OPEN

VACUUM SWITCH, NORMALLY CLOSED

LIQUID-LEVEL SWITCH, NORMALLY OPEN

LIQUID-LEVEL SWITCH, NORMALLY CLOSED

TEMPERATURE-ACTUATED SWITCH, NORMALLY OPEN

(continued)

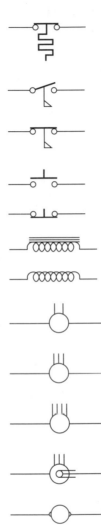

TEMPERATURE-ACTUATED SWITCH, NORMALLY CLOSED

FLOW SWITCH, NORMALLY OPEN

FLOW SWITCH, NORMALLY CLOSED

MOMENTARY-CONTACT SWITCH, NORMALLY OPEN

MOMENTARY-CONTACT SWITCH, NORMALLY CLOSED

IRON-CORE INDUCTOR

AIR-CORE INDUCTOR

SINGLE-PHASE AC MOTOR

3-PHASE, SQUIRREL CAGE MOTOR

2-PHASE, 4-WIRE MOTOR

WOUND-ROTOR, 3-PHASE MOTOR

ARMATURE

CROSSED WIRES, NOT CONNECTED

CROSSED WIRES, CONNECTED

(continued)

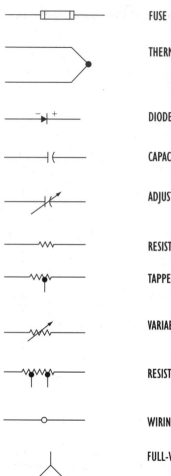

FUSE

THERMOCOUPLE

DIODE (RECTIFIRE)

CAPACITOR

ADJUSTABLE CAPACITOR

RESISTOR

TAPPED RESISTOR

VARIABLE RESISTOR

RESISTOR WITH TWO TAPS

WIRING TERMINAL

FULL-WAVE RECTIFIER

MECHANICAL INTERLOCK

MECHANICAL CONNECTION

Figure 13-1   Common symbols for motor controls.

13-1   Draw a *diagram* of a magnetic three-phase starter with one start-stop station.

See Figure 13-2.

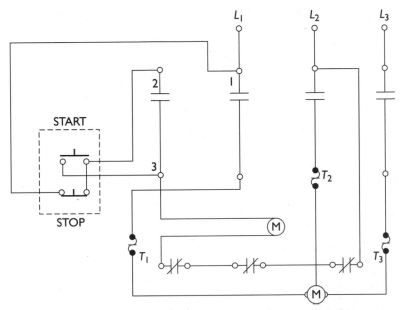

Figure 13-2   A diagrammatic representation of a magnetic three-phase starter with one start-stop station.

13-2   Draw a *schematic* of a magnetic starter with one start-stop station.

See Figure 13-3. Note that Figures 13-2 and 13-3 both accomplish the same function, but you can see why the schematic method is preferred.

13-3   What happens on the normal magnetic starter if the electricity drops off for an instant?

The magnetic starter will drop out and will stay out until restarted.

13-4   How can you restart the magnetic starter after a momentary lightning interruption?

The holding contacts (M in Figure 13-3) are shunted with a timing device that maintains the contact for a certain time period, which is ordinarily adjustable for any desired delay.

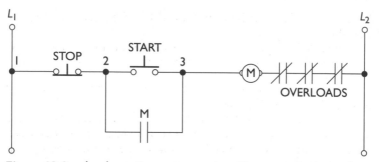

Figure 13-3 A schematic representation of a magnetic three-phase starter with one start-stop station.

13-5 In an area where a number of timing devices are used, what would happen if they were all set for the same time-delay interval?

They would all come back on at the same instant, thereby causing a tremendous current surge on the distribution system, which would cause the line breakers to trip. This situation can be avoided by setting the time delays at different intervals, so that they will be staggered when they come back into the system.

13-6 Draw a maintained-contact control, and explain how it differs from Figure 13-3.

The maintained-contact control has no voltage drop-out device; therefore, if the current is interrupted, the starter will come back on the line when electric service is restored and will function normally until another interruption (Figure 13-4).

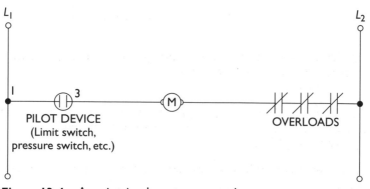

Figure 13-4 A maintained-contact control.

13-7 Name a few devices that may be used as the pilot in Figure 13-4.

Limit switch, pressure switch, thermostat, flow switch, low-water switch, and hand-operated switch.

13-8 How can you use low voltage to control a motor starter that is designed for use on high voltage?

A transformer or other source of voltage could be used if the control wires are run in the same raceways as the motor conductors. They must have insulation equal to the proper voltage rating of the highest voltage used.

13-9 Draw a schematic of a low-voltage control using a transformer. See Figure 13-5.

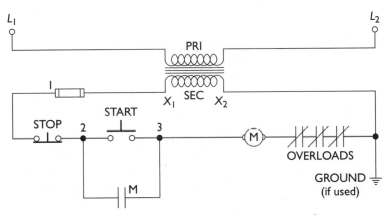

Figure 13-5 A low-voltage motor control using a transformer.

13-10 Draw a schematic of a low-voltage control using a control relay to energize the magnetic coil with full voltage.

See Figure 13-6.

13-11 Draw a schematic of a magnetic starter with one stop-start station and a pilot that burns when the motor is running.

See Figure 13-7.

13-12 Draw a schematic of a magnetic starter with three stop-start stations.

See Figure 13-8.

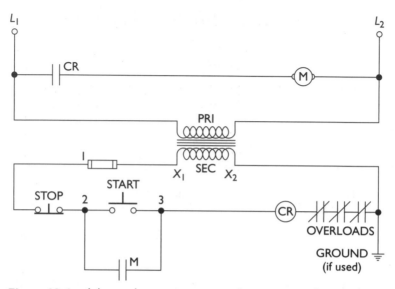

Figure 13-6 A low-voltage motor control using a transformer in conjunction with a control relay.

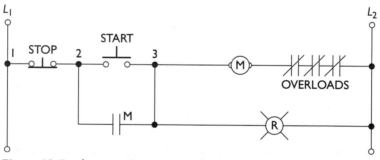

Figure 13-7 A magnetic starter with one stop-start station and a pilot lamp that burns to indicate that the motor is running.

13-13 What is jogging?
This means inching a motor, or constant starting and stopping of the motor, to move it a little at a time.

13-14 Draw a starter schematic with a jogging switch.
See Figure 13-9.

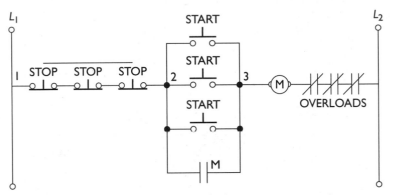

Figure 13-8 A magnetic starter with three stop-start stations.

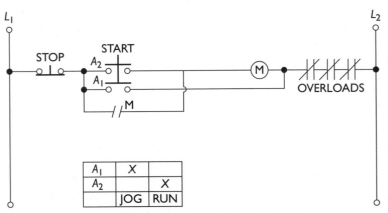

Figure 13-9 A magnetic starter with a jogging switch.

13-15 What is plugging?

This means stopping a motor by instantaneously reversing it until it stops.

13-16 Draw a starter schematic illustrating connections for plugging a motor provided with a safety latch.

See Figure 13-10.

13-17 Draw a schematic of three motors that are all started and stopped from one stop-start station so that they will all stop if one overload trips.

See Figure 13-11.

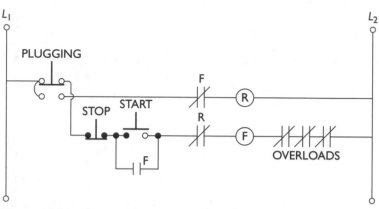

Figure 13-10 A magnetic starter with a plugging switch.

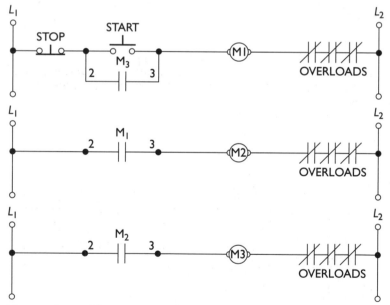

Figure 13-11 Three motors that are simultaneously controlled by one stop-start station; If one overload device trips, all three motors will stop running.

13-18 Draw a schematic of two magnetic starters controlled from one start-stop station, with a starting time delay between the two motors.
See Figure 13-12.

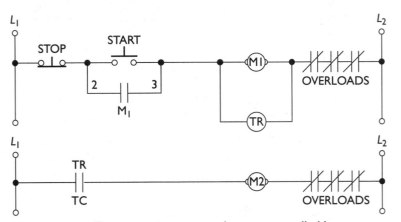

Figure 13-12 Two magnetic starters that are controlled by one start-stop station; there is a starting-time delay device between the two motors.

13-19 Draw a schematic of three separately started motors, all of which may be stopped by one master stop station or stopped if one overload trips.

See Figure 13-13.

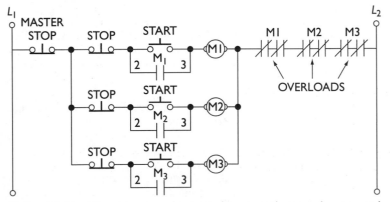

Figure 13-13 Three separately started motors that may be stopped by one master stop station or stopped if one overload device is tripped.

13-20 Draw a schematic of a starting compensator, showing start and run contacts.
See Figure 13-14.

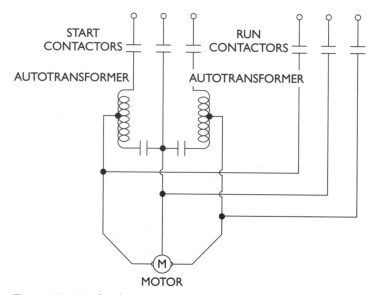

Figure 13-14 A schematic representation of a starting compensator, with start and run contacts.

13-21 Draw a diagram of a 200-ampere service with a 12½-kVA alternator as a standby; this will be a nonautomatic switchover. The diagram must be arranged so that the two sources cannot be connected together, and both must have proper overload protection.
See Figure 13-15.

13-22 Draw a schematic of a two-speed, three-phase, squirrel cage motor starter; show motor connections.
See Figure 13-16.

13-23 What is a starting compensator?
This is an ac device that consists of a built-in autotransformer, which reduces the voltage to the motor at start. After coming up to partial speed at a reduced voltage and reduced line current, the motor is connected across the line for its running position (see Figure 13-14 for the magnetic-type compensator).

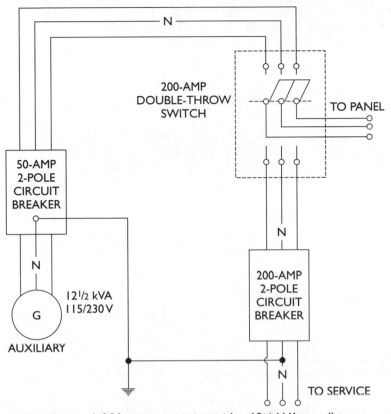

Figure 13-15   A 200-ampere service with a 12½-kVA standby alternator arranged for a nonautomatic switchover.

13-24 Draw a diagram of a manual-type starting compensator.
See Figure 13-17.

13-25 Draw a diagram of a two-speed ac motor control with push-button control.
See Figure 13-18.

13-26 Draw a diagram of the control arrangement for an ac multi-speed motor.
See Figure 13-19.

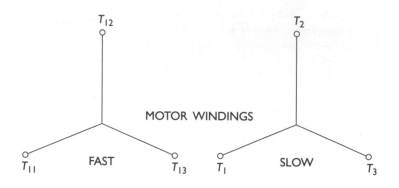

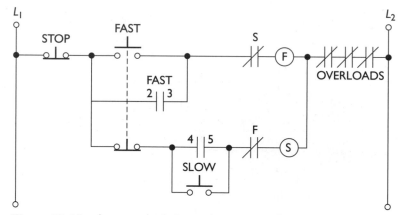

Figure 13-16 A two-speed, three-phase, squirrel cage motor starter.

13-27 Draw a wiring diagram for a dc magnetic starter with a contactor and overload relay.

See Figure 13-20.

13-28 Draw a wiring diagram of a dc shunt motor speed-regulating rheostat for starting and speed control by field control.

See Figure 13-21.

13-29 Draw a wiring diagram of a dc speed-regulating rheostat for shunt or compound-wound motors with contactors and pushbutton station control.

See Figure 13-22.

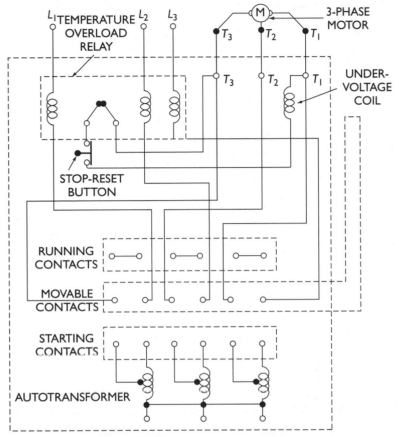

Figure 13-17 A manual-type starting compensator.

13-30 Draw a wiring diagram of a dc speed-regulating rheostat for shunt or compound-wound motors without a contactor.

See Figure 13-23.

13-31 Draw a diagram of a dc speed-regulating rheostat for shunt or compound-wound motors. Regulating duty—50 percent speed reduction by armature control and 25 percent increase by field control.

See Figure 13-24.

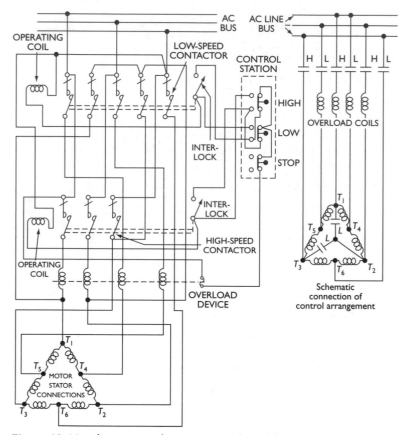

Figure 13-18 A two-speed ac motor with pushbutton control.

13-32 Draw a diagram of magnetic controller for a constant-speed dc shunt or compound-wound nonreversible motor with dynamic braking.

See Figure 13-25.

13-33 Draw a magnetically operated dc motor started with three pushbutton control stations.

See Figure 13-26.

13-34 Draw a pushbutton-operated dc motor starter showing a starting arrangement wherein the armature starting current is limited by a step-by-step resistance regulation.

See Figure 13-27.

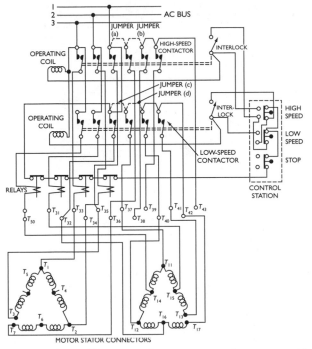

Figure 13-19 The control arrangement for an ac multispeed motor.

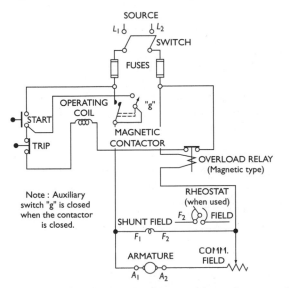

Figure 13-20 The wiring diagram for a dc magnetic starter, with a contactor and an overload relay.

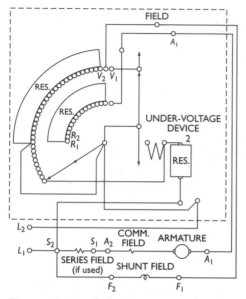

Figure 13-21   A dc shunt motor speed-regulating rheostat for starting and speed control by field control.

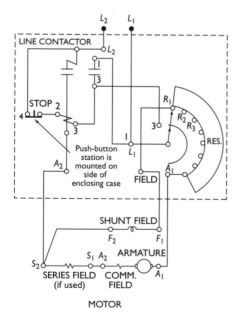

Figure 13-22   A dc speed-regulating rheostat control for shunt or compound-wound motors with contactors and pushbutton station control.

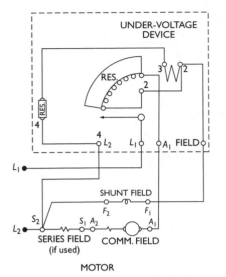

Figure 13-23 A dc speed-regulating rheostat control for shunt or compound-wound motors without a contactor.

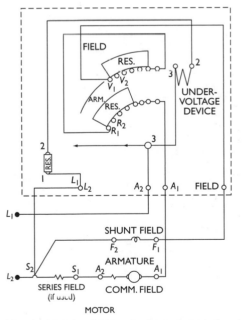

Figure 13-24 A dc speed-regulating rheostat control for shunt or compound-wound motors; regulating duty—50% speed reduction by armature control and 25% increase by field control.

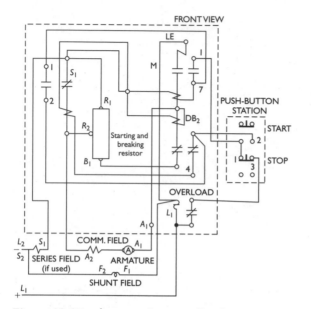

Figure 13-25 A magnetic controller for constant-speed dc shunt or compound-wound nonreversible motors with dynamic braking.

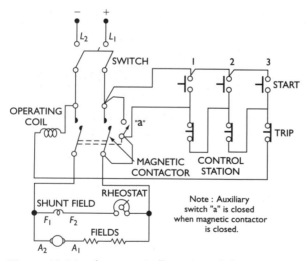

Figure 13-26 A magnetically operated dc motor starter with three push-button control stations.

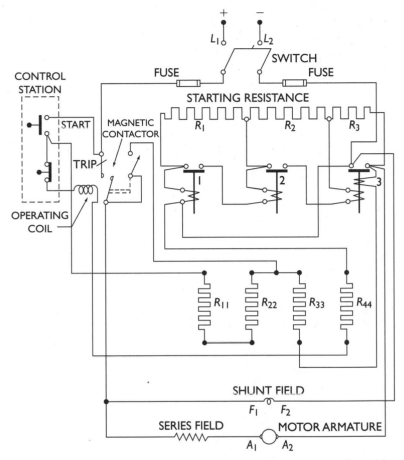

Figure 13-27 A pushbutton-operated dc motor starter in which the armature starting current is limited by a step-by-step resistance regulation.

Chapter 14

Special Occupancies and Hazardous (Classified) Locations

14-1 Where rigid conduit is used in hazardous (classified) locations, what precautions must be taken?

All threaded connections must be made wrench tight to minimize sparking when fault currents flow through the conduit system. Where this is not possible, a bonding jumper must be used.

14-2 Various atmospheric mixtures require different treatment. What is the grouping given to these various mixtures?

The characteristics of various atmospheric mixtures of hazardous gases, vapors, and dusts depend on the specific hazardous material involved. It is necessary, therefore, that equipment be approved not only for the class of location but also for the specific gas, vapor, or dust that will be present.

For purposes of testing and approval, various atmospheric mixtures have been grouped on the basis of their hazardous characteristics, and facilities have been made available for testing and approval of equipment for use in the following atmospheric groups:

Group A—Atmospheres containing substances such as acetylene.

Group B—Atmospheres containing substances such as hydrogen, or gases or vapors of equivalent hazard such as manufactured gas.

Group C—Atmospheres containing substances such as ethylene, or cyclopropane.

Group D—Atmospheres containing substances such as gasoline, hexane, naphtha, benzene, butane, propane, alcohol, acetone, benzol, lacquer solvent vapors, or natural gas.

Group E—Atmospheres containing substances such as combustible metal dust.

Group F—Atmospheres containing substances such as carbon black, coal, or coke dust.

Group G—Atmospheres containing substances such as flour, starch, or grain dust.

See *Section 500.7* in the NEC for complete coverage.

14-3 How shall equipment for hazardous (classified) locations be marked?

Approved equipment shall be marked to show the class, group, and operating temperature or temperature range, based on operation in a 40°C ambient for which it is approved. See NEC, *Section 500.8(B)*.

The temperature range, if provided, shall be indicated in identification numbers as shown in NEC, *Tables 500.8(B)* and *500.8(C)(2)*.

Exceptions are made for equipment of the nonheat-producing type, such as junction boxes, conduit, and fittings, which are not required to have a marked operating temperature.

14-4 What is a Class I location?

These are locations in which flammable gases or vapors are, or may be, present in the air in quantities sufficient to produce explosive or ignitable mixtures (see NEC, *Section 500.5(B)*).

14-5 What is a Class I, Division 1 location?

Locations in which hazardous concentrations of flammable gases or vapors exist continuously, intermittently, or periodically under normal operating conditions; in which hazardous concentrations of such gases or vapors may exist frequently because of repair or maintenance operations or because of leakage; or in which breakdown or faulty operations of equipment or processes that might release hazardous concentrations of flammable gases or vapors might also cause simultaneous failure of electrical equipment (see NEC, *Section 500.5(A)(1)*).

14-6 What is a group A atmosphere in a hazardous (classified) location?

One that contains acetylene (see NEC, *Section 500.6(A)(1)*).

14-7 What is a group C atmosphere in a hazardous (classified) location?

One that contains cyclopropane, ethylether, ethylene, or gases or vapors of equivalent hazard (see NEC, *Section 500.6(A)(3)*).

14-8   What is intrinsically safe wiring and equipment?
Equipment and wiring that is not capable of releasing sufficient electrical or thermal energy under normal or abnormal conditions to cause ignition of a specific flammable or combustible
atmospheric mixture in its most easily ignitable concentration
(see NEC, *Section 504.2*).

**14-9   What precautions must be taken when installing intrinsically
safe circuits?**
They are required to be physically separated from wiring of
all other circuits that are not intrinsically safe and from separate
intrinsically safe circuits (see NEC, *Section 504.30*).

**14-10 Must locations where pyrophoric materials are the only
materials used or handled be classified as hazardous?**
No (see NEC, *Section 500-2*).

14-11 What is pyrophoric material?
A pyrophoric material is one that is capable of igniting spon-
taneously when exposed to air.

14-12 What is a group G atmosphere?
One that contains combustible dusts such as flour, grain, or
wood (see NEC, *Section 500.6(B)(3)*).

**14-13 What is the maximum temperature that won't exceed the
ignition temperature of a specific gas or vapor that is encountered
in a hazardous (classified) location?**
See NEC, *Section 500.6* and *Table 500.8(C)(2)*.

14-14 What is a Class I, Division 2 location?
Locations in which flammable volatile liquids or flammable
gases are handled, processed, or used, but in which the hazardous liquids, vapors, or gases will normally be confined within
closed containers or closed systems from which they can escape
only in case of accidental rupture or breakdown of such containers or systems, or in case of abnormal operation of equipment; in
which hazardous concentrations of gases or vapors are normally
prevented by positive mechanical ventilation, but which might
become hazardous through failure or abnormal operation of
the ventilating equipment; or which are adjacent to a Class I,
Division 1 location and to which hazardous concentrations of
gases or vapors might occasionally be communicated unless
such communication is prevented by adequate positive-pressure

ventilation from a source of clean air, and effective safeguards against ventilation failure are provided (see NEC, *Section 500.5 (B)(2)*).

14-15 What are Class II locations?
Locations that are hazardous because of the presence of combustible dust (see NEC, *Section 500.5(C)*).

14-16 What are Class II, Division 1 locations?
Locations in which combustible dust is, or may be, in suspension in the air continuously, intermittently, or periodically under normal operating conditions in quantities sufficient to produce explosive or ignitable mixtures; where mechanical failure or abnormal operation of machinery or equipment might cause such mixtures to be produced and might also provide a source of ignition through simultaneous failure of electrical equipment, operation of protective devices, or from other causes; or in which dusts of an electrical conducting nature may be present (see NEC, *Section 500.5(C)(1)*).

14-17 What are Class II, Division 2 locations?
Locations in which combustible dust won't normally be in suspension in the air or is not likely to be thrown into suspension by the normal operation of equipment or apparatus in quantities sufficient to produce explosive or ignitable mixtures, but where deposits or accumulations of such dust may be sufficient to interfere with the safe dissipation of heat from electrical equipment and apparatus, or where such deposits or accumulations of dust on, in, or in the vicinity of electrical equipment might be ignited by arcs, sparks, or burning material from such equipment (see NEC, *Section 500.5(C)(2)*).

14-18 What are Class III locations?
Locations that are hazardous because of the pressure of easily ignitable fibers or flyings, but in which such fibers or flyings are not likely to be suspended in the air in quantities sufficient to produce ignitable mixtures (see NEC, *Section 500.5(D)*).

14-19 What are Class III, Division 1 locations?
Locations in which easily ignitable fibers or materials producing combustible flyings are handled, manufactured, or used (see NEC, *Section 500.5(D)(1)*).

14-20 What are Class III, Division 2 locations?
Locations in which easily ignitable fibers are stored or handled, except in process or manufacture of the product (see NEC, *Section 500.5(D)(2)*).

14-21 Where may meters, instruments, and relays be mounted in Class I, Division 1 locations?
Within enclosures approved for this location (see NEC, *Section 501.3(A)*).

14-22 Where may meters, instruments, and relays with make-and-break contacts be mounted in Class I, Division 2 locations?
Immersed in oil or hermetically sealed against vapor or gases (see NEC, *Section 501.3(B)*).

14-23 What type of wiring must be used in Class I, Division 1 areas?
Rigid metal conduit (threaded), intermediate metal conduit, or Type MI cable (see NEC, *Section 501.4(A)*).

14-24 What type of fittings must be used in Class I, Division 1 areas?
All boxes, fittings, and joints must be threaded. At least five full threads must be used and must be fully engaged, and all materials, including flexible connections, must be approved (explosion-proof) by the inspecting authority for these locations (see NEC, *Section 501.4(A)*).

14-25 What type of wiring must be used in Class I, Division 2 areas?
Threaded rigid metal conduit, threaded steel intermediate metal conduit, enclosed gasketed busways, or Type PLTC cable in accordance with the provisions of *Article 725*, Type MI, MC, MV, TC, or SNM cable with approved termination fittings (see NEC, *Section 501.4(B)*).

14-26 What type of fittings must be used in Class I, Division 2 areas?
Only boxes, fittings, and joints that are approved for this location (see NEC, *Section 501.4(B)*).

14-27 Why is sealing needed in conduit systems and not in Type MI cable systems in Class I areas?
Seals are provided in conduit systems to minimize the passage of gases, vapors, or fumes from one portion of the electrical

system to another through the conduit. Type MI cable is inherently constructed to prevent the passage of gases, etc., but sealing compound is used in cable-termination fittings to exclude moisture and other fluids from the cable insulation (see NEC, *Section 501.5* FPN No. 1).

14-28 When connecting conduit to switches, circuit breakers, etc., where must seals be placed?

As close as possible to the enclosure, but not more than 18 inches away (see NEC, *Section 501.5*).

14-29 When a conduit leaves a Class I, Division 1 location, where must the seal be located?

The seal must be placed at the first fitting when a conduit leaves this area; it may be on either side of the boundary (see NEC, *Section 501.5(A)*).

14-30 When a conduit leaves a Class I, Division 2 location, where must the seal be located?

The seal must be placed at the first fitting when a conduit leaves this area; it may be on either side of the boundary (see NEC, *Section 501.5(B)*).

14-31 If there is a chance that liquid will accumulate at a seal, what precautions must be taken?

An approved seal for periodic draining of any accumulation must be provided (see NEC, *Section 501.5(F)*).

14-32 What must be provided for switches, motor controllers, relays, fuses, or circuit breakers in Class I locations?

They must be provided with enclosures and must be approved for the location (see NEC, *Section 501.6*).

14-33 What is the required thickness of the sealing compound in Class I locations?

Not less than the trade size of the conduit, and in no case less than $\frac{5}{8}$-inch thickness (see NEC, *Section 501.5(C)(3)*).

14-34 Must lighting fixtures for Class I locations be approved fixtures?

They definitely must be approved for the classification in which they are used, and must be so marked. Portable lamps must also be approved for these areas (see NEC, *Section 501.9*).

14-35 Is grounding necessary in Class I locations?
Yes. It is highly important. All exposed noncurrent-carrying parts of equipment are required to be grounded. Locknuts and bushings are not adequate grounding; they must have bonding jumpers around them. Where flexible conduit is used as permitted, bonding jumpers must be provided around such conduit (see NEC, *Section 501.16*).

14-36 Is surge protection required in Class I locations?
No, but if installed, lightning-protection devices are required on all ungrounded conductors. They must be connected ahead of the service-disconnecting means and must be bonded to the raceway at the service entrance (see NEC, *Section 501.17*).

14-37 What are some of the precautions that must be taken when transformers are used in Class II, Division 1 locations?
Transformers containing a flammable liquid must be installed only in approved vaults that are constructed so that a fire cannot be communicated to the hazardous area. Transformers that don't contain a flammable liquid must also be installed in vaults or must be enclosed in tight metal housings without ventilation (see NEC, *Section 502.2(A)*).

14-38 What type of wiring must be installed in Class II, Division 1 locations?
Rigid metal conduit (threaded), threaded steel intermediate metal conduit, or Type MI cable must be used. Boxes and fittings must have threaded bosses and tight-fitting covers and must be approved for the location (see NEC, *Section 502.4(A)*).

14-39 What type of wiring must be installed in Class II, Division 2 locations?
Rigid metal conduit, intermediate metal conduit, dust-tight wireways, electrical metallic tubing, or Type MI, MC, or SNM cable. Boxes, fittings, and joints must be made to minimize the entrance of dust (see NEC, *Section 502.4(B)*).

14-40 What is needed for sealing in Class II locations?
Where a raceway provides communication between an enclosure that is required to be dust-ignition-proof and one that is not, a suitable means must be provided to prevent the entrance of dust into the dust-ignition-proof enclosure through the raceway. This means may be a permanent and effective seal, a horizontal section of raceway not less than 10 feet long, or a vertical section of

raceway not less than 5 feet long and extending downward from the dust-ignition-proof enclosure (see NEC, *Section 502.5*).

14-41 What type of fixtures must be provided for Class II locations?
The fixtures must be approved for these locations (see NEC, *Section 502.11*).

14-42 Is there a difference between Class I fixtures and Class II fixtures?
Class I fixtures must be vapor-proof and capable of withstanding and containing an explosion from within. Class I vapors usually have a higher flash point than dusts in a Class II location; therefore, the glass enclosure on Class I fixtures, while it must be heavier to withstand an explosion from within, may be smaller because of the higher flash temperatures. Class II fixtures are faced with a heat-dissipation problem because grain dusts and other types of dust have a low flashpoint temperature. Therefore, the glass enclosure on Class II fixtures must not be allowed to reach a high temperature.

14-43 Is grounding and bonding necessary in Class II locations?
Yes. All exposed noncurrent-carrying parts of equipment must be grounded. Locknuts and bushings are not adequate grounding; they must use bonding jumpers (see NEC, *Section 502.16*).

14-44 Is surge protection required in Class II locations?
If installed, lightning-protection devices of the proper type are required on all ungrounded conductors; they must be connected ahead of the service-disconnecting means and must be bonded to the raceway at the service entrance (see NEC, *Section 502.17*).

14-45 What type of wiring is required in Class III, Division 1 locations?
Threaded rigid metal conduit, threaded steel intermediate metal conduit, or Types MI, MC, SNM; boxes and fittings must have tight-fitting covers. There must not be any screw-mounting holes within the box through which sparks might escape (see NEC, *Section 503.3(A)*).

14-46 What vehicles are included under the commercial-garages classification?
These locations include places of storage, repairing, or servicing of self-propelled vehicles, including passenger automobiles,

buses, trucks, tractors, etc., in which flammable liquids or flammable gases are used for fuel or power (see NEC, *Section 511.1*).

14-47 What are the various area classifications in commercial garages?

For each floor at or above grade, the entire area is considered as a Class I, Division 2 location to a level of 18 inches above the floor. When below grade, the entire area to 18 inches above grade is considered as a Class I, Division 2 location unless the area has positive ventilation, in which case it will be 18 inches above each such floor. Any pit or depression in the floor may be considered a Class I, Division 1 location by the enforcing authority. Adjacent stockrooms, etc., must be considered as Class I, Division 2 locations unless there is a tight curb or elevation of 18 inches above the hazardous area (see NEC, *Section 511.3*).

14-48 In commercial garages, how should areas adjacent to the hazardous area be classified?

Adjacent areas that by reason of ventilation, air pressure differentials, or physical spacing are such that in the opinion of the authority enforcing this Code no hazards exist shall be classified as nonhazardous (see NEC, *Section 511.3(B)(4)*).

14-49 What material may be used in wiring spaces above Class I locations in commercial garages?

Rigid metal conduit, threaded steel intermediate metal conduit, rigid nonmetallic conduit, electrical nonmetallic tubing, or Type MI, MC, or SNM cable (see NEC, *Section 511.7(A)*).

14-50 What is required for the 125-volt, single-phase, 15- and 20-ampere receptacles and protection of personnel in a commercial garage?

The receptacles must be protected by ground fault circuit interrupters when they are used for portable lighting devices, electrical hand tools, and electrical automotive diagnostic equipment (see NEC, *Section 511.12*).

14-51 Where are seals required in commercial garages?

Wherever conduit passes from Class I, Division 2 areas to nonhazardous areas; this applies to horizontal as well as vertical boundaries of these areas (see NEC, *Section 511.9*).

14-52 What is required on equipment above the hazardous area in commercial garages?

Equipment that is less than 12 feet above the floor and that might produce sparks or particles of hot metal must be totally enclosed to prevent the escape of sparks or hot metal particles or be provided with guards or screens for the same purpose. Lighting fixtures must be located not less than 12 feet above the floor level unless protected by guards, screens, or covers (see NEC, *Section 511.7(B)(1)*).

14-53 What classification do pits or depressions of aircraft hangars require?

These are considered as Class I, Division 1 locations (see NEC, *Section 513.3(A)*).

14-54 What is the normal classification of an aircraft hangar?

The entire area of a hangar, including adjacent areas not suitably cut off from the hangar, is considered as a Class I, Division 2 location to a height of 18 inches above the floor. Areas within 5 feet, horizontally, from aircraft power plants, aircraft fuel tanks, or aircraft structures containing fuel are considered as Class I, Division 2 locations to a point extending upward from the floor to a level 5 feet above the upper surfaces of wings and engine enclosures (see NEC, *Section 513.3*).

14-55 What type of wiring is required in all aircraft-hangar locations?

The wiring must generally be in accordance with Class I, Division 1 locations (see NEC, *Section 513.4*).

14-56 What type of wiring may be used in areas that are not considered as hazardous in aircraft hangars?

Metallic raceway, Type MI cable, TC, or MC (see NEC, *Section 513.7(A)*).

14-57 How must equipment, including lighting, above the aircraft be treated?

Any equipment or lighting less than 10 feet above wings and engine enclosures must be enclosed or suitably guarded to prevent the escape of arcs, sparks, or hot metal particles (see NEC, *Section 513.7(C)*).

14-58 Is sealing required in aircraft hangars?

Yes, where wiring extends into or from hazardous areas. This regulation includes horizontal as well as vertical boundaries. Any raceways in the floor or below the floor are considered as being in the hazardous areas (see NEC, *Section 513.9*).

14-59 Is grounding required in aircraft hangars?

Yes, all raceways and all noncurrent-carrying metallic portions of fixed or portable equipment, regardless of voltage, must be grounded (see NEC, *Section 513.16*).

14-60 What is a gasoline-dispensing and service station?

This includes locations where gasoline or other volatile flammable gases are transferred to the fuel tanks (including auxiliary fuel tanks) of self-propelled vehicles. Lubritoriums, service rooms, repair rooms, offices, sales-rooms, compressor rooms, and similar locations must conform to *Article 511* of the NEC (see NEC, *Section 514.2*).

14-61 What are the hazardous areas in and around gasoline-dispensing islands?

The area within the dispenser and extending for a distance of 18 inches in all directions from the enclosure and extending upward for a height of 4 feet from the driveway; any wiring within or below this area will be considered as a Class I, Division 1 location and must be approved for this location. Any area in an outside location within 20 feet, horizontally, from the exterior enclosure of any dispensing pump is considered as a Class I, Division 2 location. Any areas within buildings not suitably cut off from this 20-foot area must also be considered as a Class I, Division 2 location. The Class I, Division 2 location within the 20-foot area around dispensing pumps extends to a depth of 18 inches below the driveway or ground level. Any area in an outside location within 10 feet, horizontally, from any tank fill-pipe must be considered as a Class I, Division 2 location, and any area within a building in the 10-foot radius must be considered the same. Electrical wiring and equipment emerging from dispensing pumps must be considered as a Class I, Division 1 location, at least to its point of emergence from this area. Any area within a 3-foot radius of the point of discharge of any tank vent pipe must be considered as a Class I, Division 1 location; and below this point, the area must be considered as a Class I, Division 2 location. (See *Table 514.3(B)(1)* of the NEC, which is very complete as to the classifications of all areas in and around service stations.)

14-62 What are the restrictions on switching circuits leading to or going through a gasoline-dispensing pump?

The switches or circuit breakers in these circuits must be able to simultaneously disconnect all conductors of the circuit, including the grounded neutral, if any (see NEC, *Section 514.11*).

14-63 Are seals required in service station locations?

Yes, approved seals must be provided in each conduit run entering or leaving a gasoline dispensing pump or other enclosure located in a Class I, Division 1 or Division 2 location when connecting conduit originates in a nonhazardous location. The first fitting after the conduit emerges from the slab or from the concrete must be a sealing fitting (see NEC, *Section 514.9*).

14-64 Is grounding required in a service station wiring system?

Yes, all metal portions of dispensing islands, all metallic raceways, and all noncurrent-carrying parts of electrical equipment must be grounded, regardless of voltage (see NEC, *Section 514.16*).

14-65 What are bulk-storage plants?

Locations where gasoline or other volatile flammable liquids are stored in tanks having an aggregate capacity of one carload or more and from which such products are distributed, usually by tank truck (see NEC, *Section 515.2*).

14-66 What are the classifications of bulk-storage plants?

Adequately ventilated indoor areas containing pumps, bleeders, withdrawal fittings, meters, and similar devices are considered as Class I, Division 2 locations in an area extending 5 feet in all directions from the exterior surfaces of such equipment; this location also extends 25 feet horizontally from any surface of the equipment and 3 feet above the floor or grade level. Areas that are not properly ventilated are required to have the same distances as above, but are considered as Class I, Division 1 locations (see NEC, *Section 515.3* and *Table 515.3*).

14-67 Is underground wiring permitted in bulk-storage plants?

Yes, underground wiring must be installed in threaded rigid metal conduit. When buried in 2 feet or more of earth, it may be installed in rigid nonmetallic conduit or duct, or in the form of approved cable. Where cable is used, it must be enclosed in rigid metal conduit from the point of lowest buried cable level to the point of connection to the above-ground raceway (see NEC, *Section 515.8*).

14-68 What are spray application, dipping, and coating processes locations?
Locations where paints, lacquers, or other flammable finishes are used regularly or are frequently applied by spraying, dipping, brushing, or other means, and where readily ignitable deposits or residues from such paints, lacquers, or finishes may occur (see NEC, *Section 516.2*).

14-69 What is the classification of spray application, dipping, and coating processes locations?
They are considered as Class I, Division 1 locations (see NEC, *Section 516.3*).

14-70 Can a direct-drive fan be used for ventilation of spray application, dipping, and coating processes areas?
No, even though an explosion-proof motor is used because the air passing over the motor contains particles of lacquers and paints. These particles will settle out on the motor, and the exterior temperature of the motor may reach a point at which these residues would become easily ignited (see NEC, *Section 516.3*).

14-71 Is there a lighting fixture approved for direct hanging in a spray-paint booth?
None have been approved because of the low flash point of residues that collect on the glass. Illumination through panels of glass or other translucent or transparent materials is permissible only where fixed lighting units are used. The panel effectively isolates the hazardous area from the area in which the lighting units are located, the lighting units are approved for their specific locations, and the panel is of material or is so protected that breakage will be unlikely. The arrangement is such that normal accumulations of hazardous residues on the surface of the panel won't be raised to a dangerous temperature by radiation or conduction from the illumination source (see NEC, *Section 516.3*).

14-72 What is a critical branch?
A subsystem of the emergency system consisting of feeders and branch circuits supplying energy to task illumination and selected receptacles serving areas and functions related to patient

care, and which can be connected to alternate power sources by one or more transfer switches. (Definition in *Section 517.2* of the NEC.)

14-73 What is a reference grounding point?

The terminal grounding bus that serves as the single focus for grounding the electrical equipment connected to an individual patient, or for grounding the metal or conductive furniture or other equipment within reach of the patient or a person who may be touching him. (Definition in *Section 517.2* of the NEC.)

14-74 What is a room bonding point?

The terminal grounding bus that serves as a single focus for grounding the patient reference grounding buses and all other metal or conductive furniture, equipment, or structural surfaces in the room.

The bus may be located in or outside the room. The room reference grounding bus and the patient reference grounding bus may be a common bus if there is only one patient grounding bus in the room. (Definition in *Section 517.2* of the NEC.)

14-75 What is the classification of a flammable anesthetics storage room?

It is considered as a Class I, Division 1 location throughout the entire area (see NEC, *Section 517.60(A)(2)*).

14-76 What classification is given to an anesthetizing location?

It is considered as a Class I, Division 1 location and this area extends upward to a height of 5 feet (see NEC, *Section 517.60 (A)(1)*). (Also see NFPA 99, Annex 2, 2.1, 2.2.)

14-77 What type of wiring is required in a flammable-anesthetics location?

Any wiring operating at more than 8 volts between conductors must conform to Class I, Division 1 location wiring specifications and must be approved for the hazardous atmospheres involved (see NEC, *Section 517.61(A)*).

14-78 What special precautions pertain to circuits in anesthetizing areas?

All circuits must be supplied by approved isolating transformers, and proper ground-detector systems approved for the location must be used (see NEC, *Section 517.61*).

Chapter 15

Grounding and Ground Testing

15-1 What is considered the best grounding electrode?

Ten feet or more of buried metallic piping (see NEC, *Section 250.52*). At least one additional made electrode must be added for grounding to varied metallic water piping.

15-2 What is the maximum allowable resistance of made electrodes (grounding)?

The resistance to ground cannot exceed 25 ohms (see NEC, *Section 250.56*).

15-3 If the resistance exceeds 25 ohms, how may this condition be corrected?

By connecting two or more electrodes in parallel (see NEC, *Section 250.56*).

15-4 When driving a ground rod, or made electrode, is the resistance near the foundation more or less than the resistance of a rod driven a few feet away from the foundation?

The resistance will usually be more when a rod is driven near the foundation. This is due to the fact that when the rod is driven several feet away, the earth is in a circle around the ground rod instead of in a semicircle, as it would be when driven close to the foundation.

15-5 Is the ground resistance more or less when a rod is driven into undisturbed soil?

When a rod is driven into undisturbed soil, the pressure is greater against the rod. This pressure lowers the ground resistance.

15-6 Must all driven grounds be tested?

Since made electrodes must, where practical, have a resistance to ground not in excess of 25 ohms, the testing of ground resistance can be required by the inspection authority.

15-7 Is the common ohmmeter (dc-type) a good instrument to use when testing ground resistance?

No.

15-8 When using a common ohmmeter for ground testing, why can the results not be relied on?

Stray ac ground currents will probably be encountered. Some dc currents may also be found in the ground; these are due to electrolysis—the battery action between the moist earth and metals in contact with it, including the grounding electrode.

15-9 What instruments should be used for testing ground resistance?

A standard ground-testing megger, a battery-operated, vibrator-type ground tester, or a transistor-oscillator-type ground tester. The resistance may also be tested with an isolating transformer in conjunction with a voltmeter and an ammeter.

15-10 When testing the ground resistance on a ground rod, should the connection to the service be disconnected from the rod?

Yes.

15-11 Why should the connection to the ground rod be removed while testing?

Because of the danger of feedback into any other ground rods, equipment, etc., which would produce an inaccurate reading.

15-12 Sketch the procedure for ground-rod testing.

See Figure 15-1.

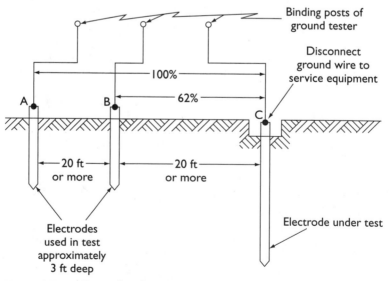

Figure 15-1 Ground-rod testing.

15-13 Why is the middle electrode (B) in Figure 15-2 set at 62 percent of the distance from A to C?

Referring to Figure 15-2, you will see that the knee of the resistance curve is at approximately 62 percent, and this point gives us the most accurate reading.

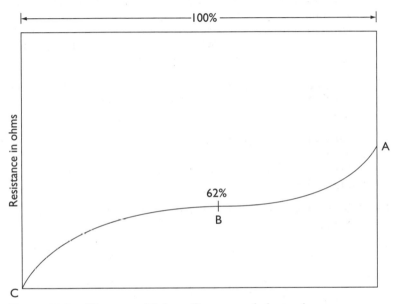

Figure 15-2   Distance of B from C measured electrode.

15-14 Sketch the procedure for testing ground resistance by the use of an isolating transformer, ammeter, and voltmeter.

See Figure 15-3.

15-15 What other method may be used to lower the resistance of a ground rod besides paralleling rods?

The use of chemicals, such as magnesium sulfate, copper sulfate, or rock salt. While the use of various chemicals and salts will lower ground resistance, it will also corrode the grounding electrode. Because of this, the use of chemicals to lower grounding resistance is often frowned upon.

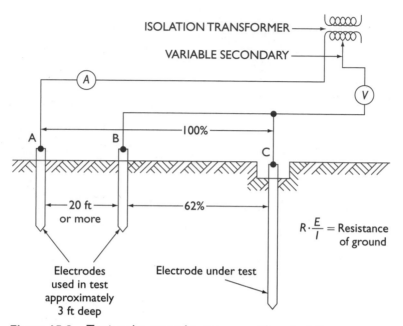

Figure 15-3 Testing the ground resistance with an isolation transformer, ammeter, and voltmeter.

15-16 Approximately what quantity of chemicals is necessary for the treatment of soil to lower its ground resistance?

The first treatment should use from 40 to 90 pounds.

15-17 How is moisture added to the chemical treatment?

The normal amount of rainfall, in most places, will provide sufficient moisture. In extremely arid areas, the problem will be different.

15-18 Sketch two common methods of adding chemicals to the ground.

See Figure 15-4.

15-19 When paralleling ground rods, will the resistance be cut in proportion to the number of rods that are paralleled?

The resistance won't be cut proportionally. For instance, two rods paralleled and spaced 5 feet apart will cut the resistance to approximately 65 percent of the original resistance.

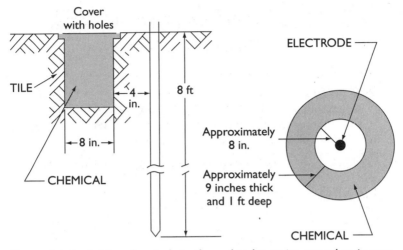

Figure 15-4 Adding chemicals to the soil to lower its ground resistance.

Three rods in parallel and spaced 5 feet apart will cut the original resistance to approximately 42 percent. Four rods in parallel and spaced 5 feet apart will cut the original resistance to approximately 30 percent.

15-20 When testing the insulation resistance, which instrument will give the better results, an ohmmeter or a megger?
The megger will give the better results because it produces a higher output voltage, which often shows up defects that the low-voltage ohmmeter cannot indicate.

15-21 What type of equipment may be used in and around swimming pools?
All electrical equipment must be approved for this type of use (see NEC, *Section 680.4*).

15-22 What is the maximum voltage allowed for underwater lighting in a swimming pool?
No lighting can be operated at more than 150 volts (see NEC, *Section 680.23(A)(4)*).

15-23 May ground-fault circuit interrupters be used on underwater lighting?
Yes (see NEC, *Section 680.23(A)(3)*).

15-24 May load-size conductors from ground-fault interrupters or transformers that supply underwater lighting be run with other electrical wiring or equipment?

No. They must be kept independent (see NEC, *Section 680.23(F)(3)*).

15-25 Must the noncurrent-carrying parts (metal) of lighting fixtures be grounded for the underwater lighting of swimming pools?

They must be grounded whether exposed or enclosed in nonconducting materials (see NEC, *Section 680.23(F)(2)*).

15-26 How must underwater fixtures be installed in swimming pools?

Only approved fixtures may be installed, and they must be installed in the outside walls of the pool in closed recesses that are adequately drained and accessible for maintenance (see NEC, *Section 680.23(C)*).

15-27 Do fixtures and fixture housings for underwater lighting have to be approved?

Yes. Approved and listed (see NEC, *Section 680.23(A)(8)*).

15-28 May galvanized rigid conduit be used to supply wet-niche fixtures?

No. Conduit of brass or other corrosion-resistance metal must be used (see NEC, *Section 680.23(B)(2)(a)*).

15-29 May ordinary isolation transformers be used for supplying underwater lighting in swimming pools?

No. The transformer and enclosure must be identified for this purpose (see NEC, *Section 680.23(A)(2)*).

15-30 How can approved equipment for swimming-pool usage be properly identified?

Look for the UL label; then look up the number and manufacturer's name in the UL listing books to ascertain whether it is listed for this type of use (see NEC, *Section 110.3(B)*).

15-31 How close to a swimming pool may an attachment-plug receptacle be installed?

Not closer than 10 feet from the inside walls of the swimming pool (see NEC, *Section 680.22(A)(2)*).

15-32 Of what material must junction boxes that supply underwater pool lights be made?

Brass or other suitable copper alloy, suitable plastic, or other approved corrosion-resistance material, if less than 4 feet from the pool perimeter and less than 8 inches above the ground or concrete (see NEC, *Section 680.24(A)(1)*).

15-33 How high or far from the pool must approved transformers be mounted?

They are to be located not less than 4 inches, measured from the inside bottom of the enclosure to the ground level or pool deck, or not less than 8 inches to the maximum pool level, whichever provides the greatest elevation; also, not less than 4 feet from the inside wall of the pool unless separated from the pool by a solid fence, wall, or other permanent barrier (see NEC, *Section 680.24(A)(2)*).

15-34 What type of metal equipment must be bonded in pool areas?

All metallic conduit, piping systems, pool-reinforcing steel, lighting fixtures, etc., must be bonded together and grounded to a common ground. This includes all metal parts of ladders, diving boards, and their supports (see NEC, *Section 680.26*).

15-35 What is considered adequate in the bonding of the reinforcing bar in concrete?

Tying the reinforcing bar together with wire, as is customarily done, is considered adequate, provided the job is well done. Welding, of course, would also be acceptable (see NEC, *Section 680.26(B)(1)*).

15-36 May the electrical equipment be grounded to a separate grounding electrode on a swimming-pool insulation?

No. The grounding must be common to the deck box or transformer ground (see NEC, *Section 680.26(C)*).

15-37 What is the minimum size of grounding conductor permitted for deck boxes on swimming pools?

The equipment grounding conductor must be No. 12 AWG or larger (see NEC, *Section 680.25(B)(1)*).

15-38 May metal raceways be relied on for grounding in swimming-pool areas?

No. An insulated equipment-grounding conductor is required (see NEC, *Section 680.25(B)*).

15-39 How must pumps, water-treating equipment, etc., in swimming-pool areas be bonded?

They must be bonded by a solid copper conductor not smaller than No. 8 AWG (see NEC, *Section 680.26(B)(4)*).

15-40 What are the clearances for service-drop conductors in swimming-pool locations?

They must be installed not less than 10 feet horizontally from the pool edge, diving structures, observation stands, towers, or platforms, and must not be installed above the swimming pool or surrounding area within the 10-foot area (see NEC, *Section 680.8*).

15-41 What type of wiring is required in theaters and assembly halls?

The wiring must be metal raceways or Type MC, MI, or ac cable, with some exceptions (see NEC, *Section 518.4* and *520.5*).

15-42 Where the assembly area is less than a 100-person capacity, may the type of wiring be altered?

Yes. The requirements of *Article 518* apply only to structures intended for the assembly of 100 persons or more. However, it is not considered good practice to use anything other than metal raceways. Also, most local ordinances or laws prohibit the use of anything other than metal raceways in public places of assembly (see NEC, *Section 518.1*).

15-43 How may the population capacity be determined in places of assembly?

Refer to the *NFPA Life Safety Code (NFPA 101)*.

15-44 What is finish rating in a place of assembly?

The time required for wood framing members to undergo a temperature rise of 121°C under fire conditions (see NEC, *Section 518.4* FPN).

15-45 What are the fixed wiring methods recognized for use in a theater or similar location?

Metal raceways, nonmetallic raceways encased in at least 2 inches of concrete, Type MI cable, or Type MC cable (see NEC, *Section 520.5*).

The answers to questions 15-46 to 15-74 may be found on pages 279–280.

15-46 In the diagram (Figure 15-5) of a 120/240-volt, three-wire circuit, load A is 10 amperes and load B is 5 amperes. Will the neutral carry any current, and, if so, how much will it carry?

(a) No current

(b) 10 amperes

(c) 15 amperes

(d) 5 amperes

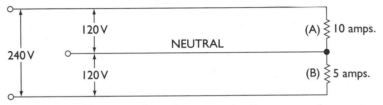

Figure 15-5 The neutral conductor carries 5 amperes of current.

15-47 In the 120/240-volt, three-wire circuit (Figure 15-6), the neutral is open at point X. The resistance of load A is 10 ohms; load B, 12 ohms; load C, 24 ohms; and load D, 20 ohms. What is the voltage drop of loads B and C, respectively?

(a) B = 160 volts; C = 80 volts

(b) B = 80 volts; C = 160 volts

(c) B = 120 volts; C = 120 volts

(d) B = 240 volts; C = 240 volts

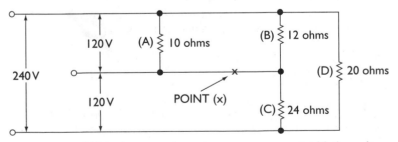

Figure 15-6 With the neutral conductor open at point X, the voltage drop of load B is 80 volts, and the voltage drop of load C is 160 volts.

15-48 If the neutral is No. 10 wire or larger, what appliance, if any, may use the neutral as an equipment ground?

(a) None

(b) Grounding-type receptacle

(c) Electric motor

(d) Electric range (except when the range is fed from a feeder panel)

15-49 When pulling current-carrying conductors into conduits, at what number of such conductors does derating of conductor current-carrying capacity begin?

(a) Four

(b) Six

(c) No derating

(d) Three

15-50 Under what condition, if any, is it possible or permissible to put overload protection in a neutral conductor?

(a) Always

(b) Never permissible

(c) Where the overload device simultaneously opens all conductors

15-51 Under what conditions may a yellow neutral be used?

(a) When extra color coding is advantageous

(b) Never

(c) On rewire work

15-52 The total load on a circuit for air conditioning only cannot exceed what percentage of the branch-circuit load?

(a) 115 percent

(b) 75 percent

(c) 80 percent

15-53 The total load of an air conditioner cannot exceed what percentage of the branch-circuit load on a circuit that also supplies lighting?

(a) 50 percent

(b) 75 percent

(c) 80 percent

15-54 Hazardous concentrations of gases or vapors are not normally present in Class I, Division 1 locations.
T F (True or False)

15-55 A location where cloth is woven should be designated as Class III, Division 1.
T F

15-56 Resistance- and reactive-type dimmers may be placed in the ungrounded conductor.
T F

15-57 A separate connection from the service drop is never acceptable as an emergency service installation.
T F

15-58 Portable, explosion-proof lighting units may be employed inside a spray booth during operation.
T F

15-59 A plug receptacle exclusively for the janitor's use may not be tapped from the emergency circuit wiring.
T F

15-60 Type NM (nonmetallic) cable may be used for under-plaster extensions.
T F

15-61 Heating cable may not be used with metal lath.
T F

15-62 Time switches need not be of the externally operated type.
T F

15-63 Heating cables in a concrete floor must be placed on at least 2-inch centers.
T F

15-64 Double-throw knife switches may be mounted either vertically or horizontally.
T F

15-65 Outlets for heavy-duty lampholders must be rated at a minimum load of

(a) 600 volt-amperes

(b) 1320 volt-amperes

(c) 2 amperes

(d) 5 amperes

15-66 Where permissible, the demand factor applied to that portion of the unbalanced neutral load in excess of 200 amperes is

(a) 40 percent

(b) 80 percent

(c) 70 percent

(d) 60 percent

15-67 Fuses must never be connected in multiple.
   T F

15-68 No. 18 copper wire may be employed for grounding a portable device used on a 20-ampere circuit.
   T F

15-69 Secondary circuits of wound-rotor induction motors require separate overcurrent protection.
   T F

15-70 A neutral conductor is sometimes smaller than the ungrounded conductors.
   T F

15-71 The largest size conductor permitted in underfloor raceways is

(a) Largest that raceway is designed for

(b) No. 0

(c) 250 MCM

(d) No. 000

15-72 The demand factor that may be applied to the neutral of an electric range is

(a) 40 percent

(b) 80 percent

(c) 70 percent

(d) 60 percent

15-73 The unprotected length of a tap conductor having a current-carrying capacity equal to one-third that of the main conductor must be no greater than

(a) 15 feet

(b) 25 feet

(c) 20 feet

(d) 10 feet

15-74 The minimum feeder allowance for show-window lighting expressed in volt-amperes per linear foot shall be

(a) 100 volt-amperes

(b) 200 volt-amperes

(c) 300 volt-amperes

(d) 500 volt-amperes

Answers to Questions 15-46 to 15-74

15-46 d.

15-47 b.

15-48 d.

15-49 a.

15-50 c.

15-51 b.

15-52 c (see NEC, *Section 440.62(B)*).

15-53 a (see NEC, *Section 440.62(C)*).

15-54 F (see NEC, *Section 500.5(B)*).

15-55 T (see NEC, *Section 500.5(D)*).

15-56 T (see NEC, *Section 520.25(B)*).

15-57 F (see NEC, *Section 700.12(D)*).

15-58 F (see NEC, *Section 516.4*).

15-59 T (see NEC, *Section 700.15*).

15-60 F (see NEC, *Section 334.12*).

15-61 F (see NEC, *Section 424.41(C)*).

15-62 T (see NEC, *Section 404.5*).

15-63 F (see NEC, *Section 426.20(B)*).

15-64 T (see NEC, *Section 404.6(B)*).

15-65 a (see NEC, *Section 220.3(B)(5)*).

15-66 c (see NEC, *Section 220.22*).

15-67 T (see NEC, *Section 240.8*, for the one exception).

15-68 T (see NEC, *Section 250.122(E)*).

15-69 F (see NEC, *Section 430.32(E)*).

15-70 T (see NEC, *Section 220.22*).

15-71 a (see NEC, *Sections 372.10* and *374.4*).

15-72 c (see NEC, *Section 220.22*).

15-73 b (see NEC, *Section 240.21(B)*).

15-74 b (see NEC, *Section 220.12*).

15-75 All 125-volt, single-phase, 15- and 20-ampere dwelling unit receptacle outlets installed above countertops within 6 feet of a kitchen sink, require ground-fault, circuit interrupter protection.
True (see NEC, *Section 210.8(A)*).

15-76 Cable assemblies with a neutral conductor sized smaller than the ungrounded conductors require no marking.
False (see NEC, *Section 210.19(A)*).

15-77 Type ac cable (BX) installed in thermal insulation is required to be a cable assembly with 90°C-rated conductors.
True (see NEC, *Section 320.80(A)*).

15-78 It is not necessary to bury lengths of rigid or intermediate metal conduit 24 inches below a driveway, street, alley, or road.
False (see NEC, *Table 300.5*).

15-79 No. 16 and 18 AWG fixture wires must be counted when sizing a junction box with a combination of other branch-circuit conductors.
True (see NEC, *Section 314.16(B)(1)*, Exception).

15-80 A second set of insulated equipment-grounding conductors, installed in a box, must be counted for box fill.
True (see NEC, *Section 314.16(B)(1)*).

15-81 Galvanized steel wires are permitted for the securing of a raceway, cable assembly, or box above a T-Bar ceiling.
True, when they provide a rigid support (see NEC, *Sections 300.11(A)* and *314.23(D)*).

15-82 A disconnecting means for an HVAC system can be installed within an air-conditioning or refrigeration equipment enclosure when a tool is required to remove the cover to gain access.
False (see NEC Definition of Accessible [equipment] and *Section 440.14*).

15-83 Identification is not required for emergency circuit wiring boxes or enclosures.
False (see NEC, *Section 700.9(A)*).

15-84 Cable trays are permitted for Type SE cable feeders in an apartment building or commercial location.
True (see NEC, *Sections 338.10(B)(4)(a)* and *334.10(4)*).

15-85 An electric baseboard heater can be installed beneath a dwelling unit receptacle outlet.
False (see NEC, *Section 110.3(B)* and *210.52* FPN).

15-86 All dwelling unit basement receptacles are required to be protected by a GFCI.
True (see NEC, *Section 210.8(A)(5)*).

15-87 Floating buildings are required to conform to the National Electrical Code.
True (see NEC, *Article 553*).

15-88 A grounded service-entrance conductor used in a three-wire, single-phase dwelling service or feeder can be reduced in size.
True (see NEC, *Section 310.15(B)(6)*).

15-89 Welding cables may not be installed in a cable tray.
False (see NEC, *Section 392.3(B)(1)*).

15-90 GFCI protection is not required for the receptacles in a repair garage.
False (see NEC, *Section 511.12*).

15-91 GFCI protection is required for a portable high-pressure sprayer.
True (see NEC, *Section 422.49*).

15-92 Electric meter socket enclosures are covered by the National Electrical Code.
True (see NEC, *Section 312.15*).

15-93 Type SE cable assemblies, used for a service, can be located next to a window, when the drip loop cannot be contacted from the sides, bottom, or top of a window.
True (see NEC, *Sections 230.9* and *338.10(A)*).

15-94 The Code recognizes the use of a raceway for enclosing a combination of systems, such as water, gas, or air lines.
False (see NEC, *Section 300.8*).

15-95 Flexible metal conduit is not permitted for use as a service-entrance raceway.
False (see NEC, *Section 230.43*).

15-96 Hospital-grade receptacles are not required within any area of a hospital other than within an inhalation anesthetizing location.
False (see NEC, *Section 517.18(B)*).

15-97 Nonmetallic sheathed cables (romex) require protection in the form of sleeves or grommets prior to being installed through the sharp edges of a metal stud partition.
True (see NEC, *Section 300.4(B)(1)*).

15-98 Ceiling paddle fans must be supported by a structural member instead of the box used at the fan outlet.
True (see NEC, *Sections 314.27(D)* and *422.18*).

15-99 Where electronic computer/data processing equipment is installed in a room, one disconnecting means can serve to disconnect the power and air-conditioning systems.
True (see NEC, *Section 645.10*).

Chapter 16

Data & Communications Wiring

While the traditional work of the electrical contractor has always been 60 Hz wiring for light and power, and while it is still the primary type of electrical work, it is no longer the only kind. Certainly, wiring for light and power won't go away, but it will comprise a far lower percentage of the work electrical installers perform.

New types of technologies are coming into almost every business and industry. Computer technology, and especially the Internet, has changed the world, and opened broad new fields for electrical construction professionals. Electronic- and computer-based technologies have grown from a small percentage of the total volume of electrical contracting in 1980 to over one third of the contractor's business today. This segment of our trade has become very important.

Likewise, the use of optical fiber for communications has been increasing steadily for many years, both in the United States and abroad. Fiber optics, which began to be used by long-distance telephone companies, has been adopted by the Cable TV industry, and is now being used widely as data cabling. Optical fiber will be the central communications technology of the twenty-first century and will provide comfortable livings for thousands of people.

A note on terms is important here: The terms *fiber optic* and *optical fiber* are used interchangeably. Optical fiber is more technically correct, and fiber optic is a more popular term. Either one is fine.

In this chapter, you will cover primarily communications installations, but will also touch upon other newer technologies, such as photovoltaics and network-powered broadband. These technologies are being found in more and more electrical tests, and it is important to give them their due. Anything in the National Electrical Code could show up in an electrical exam.

16-1   Which is the primary article of the NEC governing data cabling?
Article 800.

16-2   When data conductors enter buildings, how far must they be kept from power drops?
A minimum of 12 inches (see NEC, Section 800.10(A)(4)).

16-3   What is a "protector"?
A type of surge suppression device specifically designed for communication circuits (see NEC, *Section 800.30*).

16-4   What is the minimum size of grounding conductor for communication circuits?
No. 14 AWG (see NEC, *Section 800.40(A)(3)*).

16-5   Is firestopping required for data cables that pass through fire-resistant barriers?
Yes (see NEC, *Section 800.52(B)*).

16-6   Which standard defines structured cabling?
EIA/TIA 568.

16-7   For structured cabling, what is the horizontal distance limit between closet and desktop?
100 meters.

16-8   What is the most widely used type of data cable?
Unshielded twisted pair, also known as UTP.

16-9   Must telecommunications test equipment be listed?
No (see NEC, *Section 800.4*, Exception).

16-10 For the purposes of *Article 800*, what is meant by "block"?
A city block (see NEC, *Section 800.2*).

16-11 What types of distribution circuits don't require protectors?
Underground circuits that are unlikely to contact power circuits (see NEC, *Section 800.11(B)*).

16-12 How far must communication circuits be separated from lightning protection conductors?
6 feet, but exceptions are made when this spacing is "impractical."

16-13 What restriction is placed on the installation path of a grounding conductor for communication circuits?
That it be run in a straight line (see NEC, *Section 800.40 (A)(5)*).

16-14 May a metal power service conduit serve as the grounding electrode conductor for a communication grounding system?
Yes (see NEC, *Section 800.40(B)(1)*).

16-15 What part of a communications cable must be grounded?
The metallic sheath.

16-16 Where must a communications cable be grounded once it enters a building?
As close as possible to the point of entrance.

16-17 What option to grounding exists?
The interruption of the cable sheath with an insulating interrupting device.

16-18 When a protector is installed on a mobile home, what is the minimum size of conductor that it must be bonded with?
No. 12 AWG (see NEC, *Section 800.41(B)*).

16-19 What rating of communication cable is designed for installation in plenums?
Type CMP (see NEC, *Section 800.51(A)*).

16-20 What rating of communication cable is designed for installation in risers?
Type CMR (see NEC, *Section 800.51(B)*).

16-21 What rating of communication cable is designed for installation under carpets?
Type CMUC (see NEC, *Section 800.51(F)*).

16-22 May communications circuits share a raceway with a community antenna television system?
Yes (see NEC, *Section 800.52(A)(1)(a)(4)*).

16-23 What types of communication cables may be installed in cable trays?
Types MPP, MPR, MPG, MP, CMP, CMR, CMG, and CM (see NEC, *Section 800.52(D)*).

16-24 What is meant by the word "premises"?
A user property on the far side of the demarcation point (see NEC, *Section 820.2*).

16-25 How should CATV conductors be installed on power poles?
Below power conductors (see NEC, *Section 820.10(A)*).

16-26 What part of a coaxial cable must be grounded?
The outer conductive sheath.

16-27 What is the required clearance above a roof for a coaxial cable entering a building?

8 feet, with two exceptions.

16-28 If a grounding conductor for a coaxial system is run in metal conduit, what other requirement must be met?

Both ends of the raceway must be bonded to the grounding conductor, terminal, or electrode (see NEC, *Section 800.40 (A)(6)*).

16-29 What types of conductors are typically found in a set of network-powered broadband cables?

Coaxial, twisted-pair, fiber optic, and power conductors (see NEC, *Section 830.1* FPN No. 1).

16-30 What is a fault protective device?

A device for a network-powered broadband system that is designed to protect people and equipment in the case of faults in the system (see NEC, *Section 830.2*).

16-31 What are the required burial depths for network-powered broadband cables when installed underground?

See NEC, *Table 830.12*.

16-32 May network-powered broadband cables be installed with messenger wire?

Yes (see NEC, *Section 830.11(H)*, Exception).

16-33 What is the minimal clearance above roofs for network-powered broadband cables?

8 feet, with exceptions for garages, overhangs, and sloped roofs.

16-34 Are protectors required for network-powered broadband systems?

Yes (see NEC, *Section 830.30*).

16-35 What is the minimum size of grounding conductor that may be used for network-powered broadband systems?

No. 14 AWG (see NEC, *Section 830.40(A)(3)*).

16-36 May a power service enclosure serve as a grounding electrode for a network-powered broadband system?

Yes (see NEC, *Section 830.40(B)(1)(5)*).

16-37 If separate electrodes are used for power system grounding and for network-powered broadband grounding, what is required?

The two electrodes must be bonded with a conductor of no less than No. 6 AWG. An exception is made for bonding at mobile homes, if the mobile home is supplied with power via cord and plug, or where the mobile home has no service equipment or disconnecting means (see NEC, *Sections 830.40(D)* and *830.42(B)*).

16-38 What is the maximum permitted length for Type BLX cable?

50 feet (see NEC, *Section 830.55(D)(5)*).

16-39 What types of network-powered broadband cables may be installed in plenum areas?

Type BLP. Type BLX is also acceptable, provided that it is installed according to the requirements of *Section 300.22* (see NEC, *Section 830.55(B)*).

16-40 What are the maximum and minimum sizes of Type ITC cable conductors?

A minimum of No. 22 AWG and a maximum of No. 12 AWG are permitted for Instrumentation Tray Cables (see NEC, *Section 727.6*).

16-41 Why are optical fiber cables preferable to copper conductors for communications?

Because they can carry many times as many signals, with far less degradation of the signals. One pair of optical fiber cables can carry thousands of telephone conversations with better clarity and requiring far fewer amplifiers along its path.

16-42 What are the three basic parts of optical fiber cables?

The core, the cladding, and the outer covering.

16-43 What are the two most important things to remember when installing optical fiber cables?

First, that the pulling force should not be applied to the optical fiber, but to a "strength member." Second, that optical fiber cables cannot be sharply bent.

16-44 What is the most difficult operation of connecting optical fiber cables?

Terminating or splicing them. This can be a time-consuming and difficult process.

16-45 Optical fiber cables don't carry electricity. Why are they covered within the NEC?

For two reasons: first, because they are often installed with electrical cables and by the same installers; second, because they are dependent upon the electrical and electronic equipment to which they are connected.

16-46 Does the NEC cover all optical fiber installations?

No. It applies only to the types of installations that are covered in *Article 770* of the NEC (see NEC, *Section 770.1*).

16-47 Can optical fiber cables be installed in cable trays?

Yes (see NEC, *Section 770.52(A)*).

16-48 What types of optical fiber cables are allowed in plenums?

Types OFNP and OFPC. Also, if installed according to the rules of *Section 300.22*, OFNR, OFCR, OFN, OFNG, OFC, and OFCG may also be installed in plenums (see NEC, *Section 770.53(A)*).

16-49 What does the term photovoltaic mean?

Generating electricity from light.

16-50 What is a photovoltaic cell?

A special type of semiconductor that generates a small voltage when exposed to light, generally about one half volt.

16-51 What type of electricity do photovoltaic cells generate?

Direct current only.

16-52 What is an array?

An assembly of photovoltaic panels (see NEC, *Section 690.2*).

16-53 Why are blocking diodes required for photovoltaic systems?

To prevent battery currents from flowing backward through the photovoltaic cells when no sunlight is present.

16-54 Do photovoltaic systems require special disconnects?

Yes. A disconnecting means must be installed that will disconnect photovoltaic conductors from all other system conductors (see NEC, *Section 690.15*).

16-55 Can any recognized type of wiring be used for photovoltaic systems?

Yes (see NEC, *Section 690.31(A)*).

16-56 Must a 12-volt photovoltaic system be grounded?
No (see NEC, *Section 690.41*).

16-57 Name two types of communications signals.
Analog and digital.

16-58 In general, what type of signal is better?
Digital; it provides a much clearer and stronger signal.

16-59 What is a protector?
Protectors are surge-suppressor-type devices. They are required for many outdoor communication circuits (see NEC, *Section 800.30*).

16-60 What is the smallest grounding conductor that can be used for communication circuits?
No. 14 copper (see NEC, *Section 800.40(A)(3)*).

16-61 What steps must be taken when communication grounding conductors are installed in metal conduit?
The metal conduit must be bonded on both ends (see NEC, *Section 800.40(A)(6)*).

16-62 Should communication grounds be bonded to power system grounds?
Yes (see NEC, *Section 800.40(D)*).

16-63 How far must communication circuits be separated from power circuits?
At least 2 inches (see NEC, *Section 800.52(A)(1)*).

16-64 In what part of the NEC would you find requirements for fire alarm systems and circuits?
In *Article 760*.

16-65 What is a closed-loop system?
A special type of power distribution that is controlled jointly by signals from the controlling equipment and the utilization equipment.

16-66 What types of disconnects are required in data processing areas?
They require a disconnect for all electronic equipment in these areas. In addition, there must be another disconnect for all

HVAC equipment in these areas. These can be combined into one unit (see NEC, *Section 645.10*).

16-67 Are ventilated underfloor areas in computer rooms exempted from the special air-handling area requirements of Section 300.22?
Yes (see NEC, *Sections 300.22(D) and 645.5(D)*).

16-68 Where in the NEC are the rules governing isolated-ground systems found?
Section 250.146(D).

16-69 Conductors from a television antenna to a building must be kept how far from a 120/240 electrical service?
2 feet (see NEC, *Section 810.13*).

16-70 How far apart must supports for antenna cables be placed?
No specific distance is mentioned in the Code, just that they must be made secure (see NEC, *Section 810.12*).

16-71 What size conductor is required to connect a communications grounding electrode and a power grounding electrode?
At least No. 6 AWG copper (see NEC, *Section 800.40(D)*).

16-72 Which article of the Code covers cable television systems?
Article 820 covers all such systems that are wired in coaxial cable, as almost all cable TV systems are.

16-73 Explain the title of *Article 820*.
This article was written many years before the first cable TV systems came into effect. It was written for a very similar type of installation, community antenna television systems, and applied to cable TV systems once they came into use.

16-74 How far must coaxial cables be kept from the conductors of lighting systems?
6 feet (see NEC, *Section 820.10(F)(3)*).

16-75 What size conductor is required to connect a television system grounding electrode and a power grounding electrode?
At least No. 6 AWG copper (see NEC, *Section 820.40(D)*).

16-76 Which conductor of a coaxial cable can be grounded?
When the cable is to be grounded, it is the outer conductor that is connected to ground (see NEC, *Section 820.33*).

16-77 What types of coaxial cables may be run as risers in a building?
Types CATVP. Types CATVR, MPP, CMP, MPR, and CMR may be used if installed according to *Section 300.22* (see NEC, *Section 820.53(A)* and *Tables 820.50* and *820.53*).

16-78 Which article of the NEC covers general low voltage wiring?
Article 725.

16-79 Which section of the code describes the differences between Class 1, 2, and 3 circuits?
Section 725.2.

16-80 What is the minimum size of conductor that can be used to interconnect storage batteries?
No. 2/0 AWG copper (see NEC, *Section 690.74*).

16-81 What do the letters OFCP represent for type OFCP cable?
Optical Fiber Conductive Plenum. Refer to *Article 770.*

Index